Chordate Origins and Evolution

Chordate Origins and Evolution

The Molecular Evolutionary Road to Vertebrates

Noriyuki Satoh
Okinawa Institute of Science and Technology
Graduate University, Okinawa, Japan

AMSTERDAM • BOSTON • HEIDELBERG • LONDON
NEW YORK • OXFORD • PARIS • SAN DIEGO
SAN FRANCISCO • SINGAPORE • SYDNEY • TOKYO

Academic Press is an imprint of Elsevier

Academic Press is an imprint of Elsevier
125 London Wall, London EC2Y 5AS, United Kingdom
525 B Street, Suite 1800, San Diego, CA 92101-4495, United States
50 Hampshire Street, 5th Floor, Cambridge, MA 02139, United States
The Boulevard, Langford Lane, Kidlington, Oxford OX5 1GB, United Kingdom

Library of Congress Cataloging-in-Publication Data
A catalog record for this book is available from the Library of Congress

British Library Cataloguing-in-Publication Data
A catalogue record for this book is available from the British Library

ISBN: 978-0-12-809934-6 (Print)
ISBN: 978-0-12-803006-6 (Online)

For information on all Academic Press publications
visit our website at https://www.elsevier.com/

  Working together
to grow libraries in
developing countries

www.elsevier.com • www.bookaid.org

Publisher: Sara Tenney
Acquisition Editor: Kristi Gomez
Editorial Project Manager: Pat Gonzalez
Production Project Manager: Edward Taylor
Designer: Matthew Limbert

Typeset by TNQ Books and Journals

Contents

Preface

The origin and evolution of chordates is one of the most mysterious and intriguing phenomena in evolutionary developmental biology. Chordates are animals characterized by possession of a notochord, a dorsal neural tube, and pharyngeal gill slits. They consist of three taxa: cephalochordates, urochordates (or tunicates), and vertebrates. Chordates belong to a supraphyletic group of deuterostomes, together with echinoderms and hemichordates, and are thought to have been derived from the common ancestor(s) of deuterostomes. Vertebrates evolved by developing a body plan with the greatest complexity among metazoans.

In 1859, Charles Darwin proposed the concept of organismal evolution. Since then, the origins and evolution of chordates have been studied, discussed, and debated vigorously for more than 150 years. A huge wave of debate followed soon after Darwin's proposal because the question of chordate origins is directly or indirectly related to how vertebrates, including human beings, emerged on the Earth. Many hypotheses were proposed to explain deuterostome evolutionary scenarios and the origins of chordates.

During the 1980s, a new wave of molecular developmental biology revealed that genes encoding transcription factors and signal pathway molecules play pivotal roles in the differentiation of embryonic cells, formation of organs and tissues, and morphogenesis for construction of metazoan body plans. Shortly thereafter, another wave of evolutionary developmental biology (evo-devo) studies revealed that metazoans, from cnidarians to vertebrates, despite their diverse morphologies, utilize a very similar set of transcription factors and signal pathway molecules for body construction; these genes are sometimes collectively called a genetic toolkit.

I have been working on the developmental biology of urochordate (tunicate) ascidians for approximately 40 years. The reason why I selected ascidians as a research target was that I was interested in the mechanisms of the temporal control of development, rather than those of spatial control, on which most researchers in this field have focused. Embryonic development of ascidians is comparatively simple. Ascidians provide an excellent experimental system to study the cellular and molecular mechanisms of temporal control of embryogenesis. After the discovery that embryonic processes are controlled not by a single clock, but by multiple clocks, some associated with zygote cytoplasm and some with embryonic cell nuclei, I became much interested in the notochord of ascidian larvae. Because the notochord is the most prominent feature of chordates, elucidation of mechanisms involved in notochord formation might advance our understanding of not only ontogenetic mechanisms, but also phylogenetic mechanisms. My laboratory at Kyoto University discovered that *Brachyury*, which encodes a member of the T-box transcription factor family, plays a pivotal role in notochord formation in ascidians and that this gene is also expressed in the archenteron invagination region of nonchordate deuterostome embryos that do not form a notochord.

Understanding of the origins and evolution of chordates advanced relatively little through the late 1980s. However, since the early 1990s, molecular developmental biological techniques allowed these questions to be addressed by evo-devo investigations. Many previous opinions, hypotheses, or theories were reexamined, and/or new hypotheses were proposed for the evolutionary scenarios of deuterostome and chordate origins. One of most discussed hypotheses relates to the inversion of the dorsoventral axis of bilaterians. This theory was rooted in the comparative anatomy of annelids, arthropods, and vertebrates in the early 19th century. It proposes that most of the components for bilaterian body construction had already evolved in their common ancestors and that vertebrates developed by inverting the dorsoventral axis compared with that of annelids. The hypothesis has received great support from molecular developmental biology.

On many occasions after hearing lectures on this story at international meetings and upon reading new manuscripts dealing with this topic, I always thought that even if the dorsoventral axis is inverted, that alone cannot explain the occurrence of the notochord in chordates. Because I had been engaged in evo-devo studies of notochord formation for a decade, I felt that the significance of this structure was being ignored. I conceived this idea in about 2005, and in 2008 I proposed the "aboral-dorsalization hypothesis" of chordate origins, in which I emphasized that the emergence of fish-like larvae having a notochord was a key developmental event to understand chordate origins. Because my explanation of the hypothesis was clumsy and a little bit odd, my hypothesis did not attract much attention of researchers in this field, and in fact it was essentially ignored.

I wished to reconsider this problem more clearly and to better summarize my thinking about this problem. This is the reason why I began preparations for this book. However, the subject of chordate origins and evolution includes dramatic and drastic changes in embryogenesis and adult morphology. The question is so huge that it is far beyond the reach of a single person. I have spent the last 5 months preparing for this endeavor. For the first 3 months, I struggled, mainly because I myself had no clear idea how chordates originated. I cannot support the dorsoventral inversion hypothesis, but then how can one explain mechanisms of chordate origins? I had no clear answer. Several ideas came to mind as images, but most of them had weak points of one sort or another. Therefore, I decided instead to prepare this book in more of an essay style, in which I could express my thoughts more freely, not so constrained by previous hypotheses.

Thereafter, I felt very much liberated from various pressures. Therefore, this book is not always an orthodox biological treatise, replete with cited references on related subjects. Some topics have been intentionally ignored and others have been preferentially and repeatedly cited to explain or promote my ideas about deuterostomes and the origins and evolution of chordates.

At the risk of being repetitious, the question of chordate origins and evolution is so huge and difficult that it may never be fully resolved. There are many potential interpretations of this problem. I anticipate that the subjects discussed in this book will provoke many responses, most negative, and some positive. However, the main reason that I wrote this book was and is to promote further discussion of this subject from various points of view. It is my hope that I will have the opportunity to revise it in the future, including issues and discussions raised by this first version.

I especially thank Dr. Steven D. Aird, who carefully checked every sentence in the book, word by word. His useful comments and suggestions on the entire content of the book were also so helpful. Truthfully, without his great help, this book never would have been completed in such a short period of time. The great technical assistance of Kanako Hisata in the preparation of figures and tables is also acknowledged.

Most of the ideas discussed in this book resulted from discussions that occurred during the Okinawa Winter Course for Evolution of Complexity, held at Okinawa Institute of Science and Technology Graduate University (OIST) in 2009–2014. I thank Drs. Chris Lowe, Jr-Kay Yu, Yi-Hsien Su, John Gerhart, Daniel Rokhsar, Michael Levine, Robb Krumlauf, Nipam Patel, Chris Amemiya, and other members of the teaching staff of the Winter Course for their discussions and suggestions. Thanks also to Oleg Simakov, Yuuri Yasuoka, Yi-Jyun Luo, and members of the Marine Genomics Unit for their comments and suggestions.

Most of my work on this book was done at the Marine Biological Laboratory of Hiroshima University. I appreciate Kunifumi Tagawa, Tatsuya Ueki, and other staff members of the laboratory for providing me with a very quiet atmosphere for thinking and writing. The technical assistance of Shoko Tanahara is also acknowledged. Finally, I thank my wife, Mikako Satoh, for her daily support and constructive comments on my work style throughout my career.

Noriyuki Satoh

Chapter 1

Deuterostomes and Chordates

This book discusses the origin and evolution of chordates. Chordates are animals characterized by the possession of a notochord, a dorsal neural tube, somites, pharyngeal gills, an endostyle, and a postanal tail. Chordates comprise three major taxa: cephalochordates (lancelets), urochordates or tunicates (including ascidians), and vertebrates (including humans). Chordates, together with echinoderms (sea stars and sea urchins) and hemichordates (acorn worms), are deuterostomes.

1.1 A BRIEF BACKGROUND

The Earth is believed to have been formed approximately 4600 million years ago. Since then it has fostered a diverse array of life forms. Except for viruses, the status of which is uncertain, living things are categorized into three domains: Bacteria, Archaea, and Eukaryota. A recent hypothesis based on molecular data suggests that eukaryotes may be subdivided into six major taxonomic ranks, including Opisthokonta, Amoebozoa, Archaeplastida, Chromalveolata, Rhizaria, and Excavata. Metazoans or multicellular animals are members of the Opisthokonta, and they are further categorized, based on their body plans, into 34–37 phyla, ranging from sponges to vertebrates.

Vertebrates are the metazoans that manifest the greatest morphological and physiological complexity. Several key embryological events are thought to have promoted evolution of the vertebrates (Fig. 1.1; eg, Nielsen, 2012). These include

1. Multicellularization, which caused single-cell organisms (choanoflagellate-like eukaryotes) to give rise to multicellular animals.
2. Evolution of diploblasts, including cnidarians. They are radially symmetrical and possess two germ layers, ectoderm and endoderm, but they lack mesoderm.
3. Evolution of triploblasts. These organisms are mostly bilaterally symmetrical and are composed of three germ layers.
4. Diversification of protostomes and deuterostomes. Protostomes include spiralians and ecdysozoans, the former being represented by molluscs and annelids and the latter by arthropods.
5. Evolution of ambulacrarians from the deuterostome ancestor(s).
6. Evolution of chordates from deuterostome ancestor(s).
7. Evolution of vertebrates among chordates.

Chordate Origins and Evolution. http://dx.doi.org/10.1016/B978-0-12-802996-1.00001-8

1

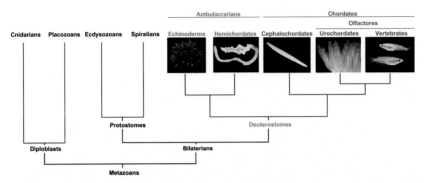

FIGURE 1.1   **Phylogenic relationships of metazoans.** Metazoans are categorized into two major groups: diploblasts (radiates) and triploblasts (bilaterians). Bilaterians in turn are subdivided into protostomes and deuterostomes. Deuterostomes comprise echinoderms, hemichordates, cephalochordates, urochordates (tunicates), and vertebrates, the first two being categorized as ambulacrarians and the last three as chordates. Therefore chordates originated from the common ancestor of deuterostomes.

This book addresses the latter three events mentioned here, and the entire evolutionary history of metazoans occupies a relatively small portion of the volume. Nonetheless, the origin and evolution of chordates culminating in the vertebrates include dramatic changes in embryogenesis and adult morphology and physiology, the dynamics of which are comparable in magnitude to those of all of the other aforementioned events combined.

1.2   DEUTEROSTOMES AND CHORDATES

The superphyletic metazoan taxon, Deuterostomia, includes the Echinodermata, Hemichordata, Cephalochordata, Tunicata (Urochordata), and Vertebrata (eg, Brusca and Brusca, 2003; Ruppert et al., 2004; Nielsen, 2012). Although classical taxonomy, based on embryological criteria, also once included chaetognaths (arrow worms) and pogonophorans (tube worms; eg, Margulis and Schwartz, 1998), recent molecular phylogenetics robustly supports their classification as protostomes (eg, Dunn et al., 2008, 2014; Philippe et al., 2009). Xenoturbellid worms are still enigmatic (Telford, 2008). These animals resemble acoelomorphs (acoel flatworms and nematodermatids) and have been grouped with them in a clade called the Xenacoelomorpha. Some molecular analyses have suggested that *Xenoturbella* and its relatives are ambulacrarians, and therefore, deuterostomes (Bourlat et al., 2006; Philippe et al., 2011), whereas other studies opine that acoelomorphs diverged from the bilaterian stem before the protostome–deuterostome split (Hejnol and Martindale, 2008; Simakov et al., 2015). In any event, if either or both of these phyla are truly deuterostomes, then their simple body plans represent a secondary loss of complexity, and they are unlikely to offer much insight into chordate origins (Chapter 4).

TABLE 1.1 Diagnosis of the Deuterostomes

Embryological Features

Radial, indeterminate cleavage

Blastopore does not form mouth, which is secondary

Mesoderm forms from infolding of gut wall

Enterocoelic coelom

Dipleurula-type larva, prototroch around the mouth

Adult Features

Tripartite body

Intraepidermal nervous system

Mesodermal skeleton

Monociliate cells?

1.2.1 Deuterostomes

Deuterostomes were first defined by Grobben (1908) as animals that share the ancestral character of deuterostomy, in which the blastopore develops into the anus and the mouth develops from a secondary opening. Radial cleavage, indeterminate cleavage of early embryos (in which blastomeres retain totipotency during early embryogenesis), dipleurula-type larvae, and enterocoely (the pouching out of mesoderm from the archenteron wall) are also distinguishing features (Table 1.1). The adult body is characterized by its triploblastic composition, with a nervous system derived from the ectoderm and a mesodermal skeleton in two taxa (Table 1.1). These contrast with the shared ancestral character of protostomy, a mouth derived from the blastopore, spiral cleavage, deterministic embryogenesis (in which cell fates are determined very early in embryogenesis), schizocoelic coelom (the splitting of mesoderm from the archenteron wall), and trochophore larvae. Although these criteria have been challenged by recent evolutionary developmental biology, they have been conventionally used as diagnostic criteria to distinguish the two major taxa of bilaterians (Schaeffer, 1987; Willmer, 1990; Gee, 1996; Hall, 1999; Brusca and Brusca, 2003; Ruppert et al., 2004; Nielsen, 2012).

1.2.2 Ambulacraria

Echinoderms and hemichordates have recently been grouped in the "Ambulacraria." Echinoderms, especially sea urchins, have provided an excellent experimental system for embryology because of their ready availability of ripe gametes. Although they show unique pentameric adult symmetry and a hard exoskeleton, the similarity of their early embryogenesis and larvae to those of hemichordates evinces the phylogenic affinity of the two taxa. Kowalevsky (1866b) and Bateson (1886) suggested that the gill slits of acorn worms and chordates are homologous, leading Bateson to christen acorn worms as "hemichordates" to emphasize their affinity with chordates. Around the same time, Metchnikoff (1881) noted similarities between the larval forms of hemichordates and echinoderms and combined these phyla into the Ambulacraria, a surprising grouping at that time, but now strongly supported by molecular phylogenetics. This unity of echinoderms and hemichordates is a prime example of the power of comparative embryology combined with recent molecular systematics.

1.2.3 Chordates

Chordates are easily distinguished from other deuterostomes by characteristic features of their body plans. The most distinctive of these are related to motility. Paired caudal muscles exert force on the notochord, a flexible skeletal rod made of disc-shaped vacuolated cells. During swimming, the notochord provides elastic recoil for the muscular undulations of the postanal tail. Chordates also possess a unique, tubular, central nervous system positioned along the dorsal midline. A series of chordate features are associated with filter feeding. These include an organ called an endostyle, associated with the pharynx. The endostyle secretes mucous to trap food and to conduct it to the gut. Pairs of pharyngeal gill slits facilitate water movement through the anterior gut.

Multicellular animals are often divided into vertebrates and invertebrates. Historically, this classification dates back to c.500 BC. During the ancient Hindu era, Charaka distinguished between the *Jarayuja* (invertebrates) and *Anadaja* (vertebrates). In the ancient Greek era, Aristotle (c.300 BC) recognized animals with blood (*Enaima*, or vertebrates) and those without (*Anaima*, or invertebrates). This recognition persisted even until Linnaeus (1766–67). It was Lamarck (1794) who first explicitly proposed the division of animals based upon the presence or absence of vertebrae, *Animaux vertébrés* and *Animaux invertébrés*, in place of *Enaima* and *Anaima*.

Aristotle had already recognized solitary ascidians as Tethyon around 330 BC. Carolus Linnaeus was a botanist who devised a system for naming plants and animals. In his book, *Systema Naturae* (12th ed., vol. 1, 1766–67), ascidians were grouped with molluscs. Following anatomical investigations of ascidians by Cuvier (1815) and others, Lamarck (1816) recognized these as Tunicata, animals enclosed with a tunic (tunica, in Latin, meaning a garment). On the other hand, cephalochordates (lancelets) were first described in the mid- to late 18th century as molluscs. Although Yarrell (1836) had already noticed that lancelets have an axial rod, calling it "a lengthened internal vertebral column, although in a soft cartilaginous state," it was Alexander Kowalevsky's discovery that tunicates and lancelets possess notochords and dorsal neural tubes during embryogenesis, which indicated that they are close relatives of vertebrates (Kowalevsky, 1866a, 1867).

The term *Vertebrata* was first coined by Ernst Haeckel in 1866, in which lancelets were assigned to the Class Acrania of the Subphylum Leptocardia and all remaining vertebrates were placed in the Subphylum Pachycardia (ie, Crania). At that time, the Tunicata was still included with bryozoans in the Subphylum Himatega of the Phylum Mollusca. Following Kowalevsky's discovery of the notochord in ascidian larvae (Kowalevsky, 1866a), Haeckel (1874a,b) moved the Tunicata from the Phylum Mollusca to the Phylum Vermes, because he thought that tunicates were close relatives of vertebrates. The Vermes also contained enteropneusts (acorn worms) because Bateson (1886) regarded the stomochord or buccal diverticulum of enteropneusts as a notochord and classified this animal as a member of the Hemichordata ("half-chord"), the fourth subphylum of the Chordata (see next section).

Haeckel coined the name *Chordonia* (ie, Chordata) for a hypothetical common ancestor of the Tunicata and the Vertebrata (including lancelets) by emphasizing the notochord as their most significant shared diagnostic character. Later, Haeckel (1894) redefined Chordonia to include the Tunicata and the Vertebrata themselves. In London, Lankester (1877) gave subphylum status to the Urochordata, the Cephalochordata, and the Craniata, altogether comprising the Phylum Vertebrata. This constituted the first conception of the modern Phylum Chordata. Balfour (1880) renamed Lankester's Vertebrata "Chordata" and called the Craniata "Vertebrata." This system has been retained for more than a century because of robustness of the shared character set (notochord, dorsal nerve cord, and pharyngeal slits) that Lankester defined.

1.2.4 Olfactores

Within the chordate clade, cephalochordates apparently diverged first whereas urochordates and vertebrates form a sister group, which is sometimes called the Olfactores (Jefferies, 1991). This name emphasizes extensive, shared pharyngeal modification leading to the formation of new structures, not found in cephalochordates.

1.3 DEUTEROSTOME PHYLA

The following are brief descriptions of the diagnostic features of the five deuterostome phyla (Brusca and Brusca, 2003; Ruppert et al., 2004; Nielsen, 2012). These provide basic knowledge for discussions of chordate origins in subsequent chapters.

1.3.1 Echinoderms

The Phylum Echinodermata (Greek *echinos*, "spiny"; *derma*, "skin") contains approximately 7000 living species with five distinct classes, including the Crinoidea (sea lilies and feather stars), Asteroidea (sea stars), Ophiuroidea (brittle stars), Echinoidea (sea urchins and sand dollars), and Holothuroidea (sea cucumbers; Fig. 1.2A). In addition, there are approximately 13,000 fossil species (Chapter 3).

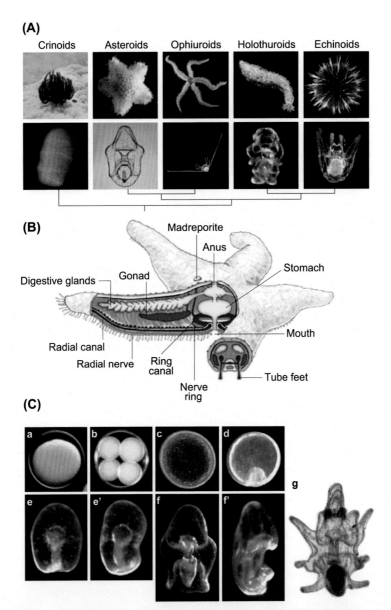

FIGURE 1.2 Echinoderms. (A) Five extant classes of echinoderms: from left to right, crinoids (sea lilies), asteroids (sea stars), ophiuroids (brittle stars), holothuroids (sea cucumbers), and echinoids (sea urchins) (upper panels). All echinoderm adults are pentaradially symmetrical, a considerable modification of the ancestral bilaterian body plan. However, their larvae are bilaterally symmetric (lower panels). Phylogenetic relationships of the five classes are shown at the bottom. *(From Lowe, C.J., Clarke, D.N., Medeiros, D.M., Rokhsar, D.S., Gerhart, J., 2015. The deuterostome context of chordate origins. Nature 520, 456–465.)* (B) Echinoderms are characterized by a conserved body plan, most clearly represented by the Asteroidea. Here a diagram of an asteroid with cutaways to show internal anatomy illustrates the major features, including the mesoderm-derived

Echinoderms are benthic, marine organisms that constitute one of the best-defined animal phyla. First, adult echinoderms are the only animals with pentameric, radial symmetry (Fig. 1.2A, upper), although their larvae are bilateral (Fig. 1.2A, lower). Second, they possess a unique calcareous endoskeleton arising from mesodermal tissue and composed of separate plates or ossicles. Each plate originates as a single mesh-like structure called a stereo, the interstices of which are filled with living tissue (stroma). Third, their left mesocoel (hydrocoel) constitutes a water vascular system composed of a complex series of fluid-filled canals, usually evident externally as muscular podia. In addition, echinoderms contain unusual types of connective tissue, the stiffness of which is modulated by neuropeptides (Birenheide et al., 1998; Santos et al., 2005).

Prosomes, mesosomes, and metasomes are unrecognizable externally, but the development of compartments from coelomic pouches of bilateral larvae clearly reveals a body organization with three distinct coelomic cavities: protocoel, mesocoel, and metacoel (called archimery). The nervous system is not centralized and usually consists of a nerve net and radial nerves. The lack of a brain-like structure results in a nervous system that is complicated and unusual. There are ring nerves around the esophagus and radial nerves along the ambulacra (Fig. 1.2B). These nervous systems possess an ectodermal (ectoneural) nerve and a mesodermal (hyponeural) nerve separated by a basement membrane. The former is mainly sensory whereas the latter is motor. Gonads are formed from mesodermal elements of the metacoel. Gap junctions have not been observed in any species. Echinoderms lack excretory organs. Circulation depends upon a hemal system derived from coelomic cavities and sinuses.

Most echinoderms are dioecious, and development is usually indirect. Their embryology is fundamentally deuterostomous, with radial cleavage, a hollow blastula, gastrulation by invagination of endodermal cells, endodermally derived mesoderm, enterocoely, and a blastopore that forms the anus (Fig. 1.2C; Wray, 1997; Davidson, 2006; McClay, 2011). The digestive system is complete, but it has become secondarily incomplete or lost in some species.

1.3.2 Hemichordates

The Phylum Hemichordata comprises two classes: Enteropneusta and Pterobranchia (Fig. 1.3A). All hemichordates are marine, benthic animals. Enteropneusts (acorn worms) are solitary, elongate, vermiform animals (Fig. 1.3A).

◀ water vascular system, a hydraulic system that drives the distinctive tube feet used for feeding and locomotion, five radial nerves that run along each arm/ambulacrum linked by a nerve ring, and the mesoderm-derived skeleton. *(From Lowe, C.J., Clarke, D.N., Medeiros, D.M., Rokhsar, D.S., Gerhart, J., 2015. The deuterostome context of chordate origins. Nature 520, 456–465.)* (C) Embryogenesis and larvae of the crown-of-thorns starfish, *Acanthaster planci*. (a) Fertilized egg, (b) four-cell embryo, (c) blastula, (d) early gastrula, (e, e′) late gastrula, (e) front view and (e′) side view, (f, f′) bipinnaria larva, (f) front view and (f′) side view, and (g) brachiolaria larva, front view. *(Courtesy of Keita Ikegami.)*

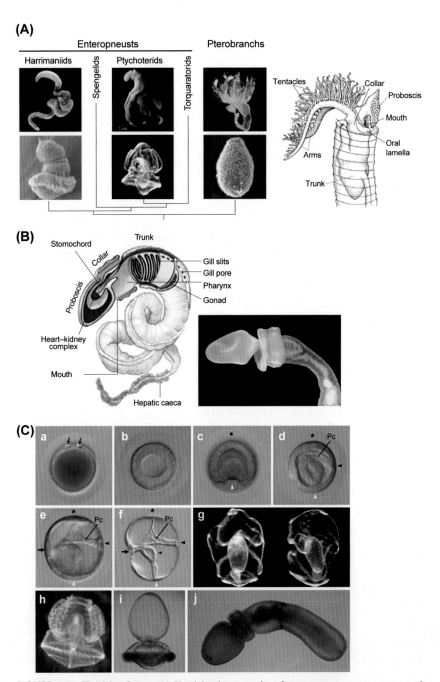

FIGURE 1.3 Hemichordates. (A) Hemichordates consist of two groups—enteropneusts and pterobranchs—and their phylogeny is shown at the bottom. Pterobranchs, represented here by *Cephalodiscus*, are small, largely colonial animals that live within a protective secreted fibrous tube and use a ciliated lophophore for filter feeding (right). Enteropneusts, or acorn worms, are

They generally live buried in soft sediments in the intertidal zone, but several deep-sea species have been recognized (Halanych et al., 2013). They feed by a combination of deposit and filter feeding. Pterobranchs are largely colonial. Zooids are small, with mesocoelic extensions into the arms and tentacles, as in lophophorates (Fig. 1.3A; Sato et al., 2008). Ciliated lophophores are used for filter feeding (Fig. 1.3A). These two groups represent highly divergent classes within the phylum, with different lifestyles, but they are united by their tripartite body plan, which includes the prosome/proboscis, the mesosome/collar, and the metasome/trunk (Fig. 1.3B; Hyman, 1959).

The enteropneust proboscis is muscular and is used for digging and feeding. A complete gut terminates at the anus at the posterior end of the long trunk. They possess a preoral gut (buccal) diverticulum. The anterior gut of hemichordates is perforated dorsolaterally with a series of ciliated gill slits supported by gill bars composed of acellular collagen secreted by the endoderm (Fig. 1.3B). In some species, a distinct genital region houses the gonads and the hepatic region is also visible with distinct coloration (Fig. 1.3B). Hemichordates have well-developed, open circulatory systems and unique excretory structures called glomeruli. Circular and longitudinal muscles are present in the body wall of the proboscis and collar of enteropneusts whereas pterobranchs have only longitudinal muscles. In the enteropneust proboscis, a basement membrane produces a rigid plate called the proboscis skeleton.

The hemichordate nervous system is characterized by two distinctive features (see Fig. 9.1): a broad epithelial plexus, particularly prominent in proboscis ectoderm, and two nerve cords. The ventral cord extends the length of the trunk whereas the dorsal cord runs from the base of the proboscis down the length of the animal, connected to the ventral cord by lateral nerve rings. Both are superficial condensations of the nerve plexus, except for a short length

◀ solitary, burrowing worms that feed using a combination of deposit and filter feeding. The harrimaniid, *Saccoglossus kowalevskii*, is a direct developer that has been used for many developmental studies (left). The ptychoderid, *Ptychodera flava*, is an indirect developer (middle). *(From Lowe, C.J., Clarke, D.N., Medeiros, D.M., Rokhsar, D.S., Gerhart, J., 2015. The deuterostome context of chordate origins. Nature 520, 456–465.)* (B) Both groups of hemichordates are united by their tripartite body plan, which includes a proboscis, a collar, and a trunk (as shown by a spengelid enteropneust). The proboscis is used for digging and feeding and contains the gut diverticulum, called the stomochord, that supports a heart–kidney complex. The mouth opens ventrally into the pharynx within the collar region, and the anterior trunk is perforated by a series of dorsolateral gill slits. *(Left: from Lowe, C.J., Clarke, D.N., Medeiros, D.M., Rokhsar, D.S., Gerhart, J., 2015. The deuterostome context of chordate origins. Nature 520, 456–465; right: courtesy of Kunifumi Tagawa.)* (C) Embryogenesis of *Ptychodera flava*. (a) Fertilized egg, *red arrows* indicating polar bodies; (b) blastula; (c) early gastrula, *white arrowhead* indicating the position of the blastopore and *asterisk* the animal pole; (d) mid-gastrula; Pc, hydrocoel, *arrowhead* indicating the hydropore; (e) late gastrula, *arrow* indicating future mouth-opening site; (f) early tornaria larva, *red arrowhead* indicating tripartite gut; (g) 2-month-old tornaria larva; (h) 6-month-old tornaria larva; (i) larva immediately before metamorphosis; and (j) juvenile 3 days after metamorphosis. *(a–f: Courtesy of Jr-Kai Yu; Lin, et al., 2016. Journal of Experimental Zoology Part B 326, 47–60) and g–j: courtesy of Kunifumi Tagawa.)*

spanning the collar, where the cord is internalized into a tube with a prominent lumen, in some species. It is formed by a developmental process that resembles chordate neurulation (Chapter 9; Morgan, 1894; Kaul and Stach, 2010; Miyamoto and Wada, 2013).

Hemichordates are dioecious. Externally fertilized eggs develop either indirectly (*Balanoglossus* and *Ptychodera*) or directly (*Saccoglossus*). Cleavage is radial and holoblastic, and blastomeres are more or less equal (Fig. 1.3C). Indirectly developing embryos become typical tornaria larvae, with a planktonic period of several months before metamorphosis (Fig. 1.3C). In contrast, direct developers adopt the adult body plan within days. Despite morphological differences between these strategies, striking similarities include formation of an unpaired anterior coelom and two pairs of posterior coeloms (Röttinger and Lowe, 2012), very similar to echinoderm larvae (Fig. 1.2A). Body cavities are formed by enterocoely. Asexual reproduction is common in pterobranchs, but their embryogenesis remains little known.

In relation to chordate mesoderm formation, the protocoel, stomochord, and the heart–kidney complex that are formed in the proboscis have attracted much interest, especially the stomochord, a diverticulum of the anterior gut that extends into the posterior proboscis (Fig. 1.3B). The cells of the stomochord are vacuolated and surrounded by a sheath, similar in cellular organization to a notochord. Possible homology of the stomochord and the notochord will be discussed in Chapter 9.

1.3.3 Cephalochordates

The Cephalochordata (lancelets) is a small phylum that contains only approximately 35 marine, fish-like creatures, among which *Branchiostoma* (amphioxus) is the best known (Fig. 1.4A). Adults are thin, fusiform, filter feeders, and most species live buried in coarse sand, but they swim very rapidly for dispersal and mating. The adult cephalochordate body is covered with an epidermis of simple columnar epithelium with an underlying thin, connective tissue dermis. The body possesses conspicuous myotomes, arranged longitudinally and dorsolaterally (Fig. 1.4A and B). The notochord persists in adults, providing structural support for the body. It consists of discoidal lamellae composed of muscle cells (Fig. 1.4B).

Cephalochordates are ciliary-mucous suspension feeders (Fig. 1.4B). Water is driven into the mouth and pharynx and out through the pharyngeal gill slits into a surrounding atrium. The oral and preoral regions of lancelets have many structures specialized for feeding and environmental sensing, including an oral hood, buccal cirri, velar tentacles, a wheel organ, and Hatschek's pit (Fig. 1.4B). The ventral surface of the pharynx bears the endostyle or hypobranchial groove (Fig. 1.4B). The gut extends posteriorly as an elongate intestine and empties through the anus in front of the caudal fin. The postanal tail is also one of the features of chordates. Near the junction of the pharynx and esophagus an anteriorly

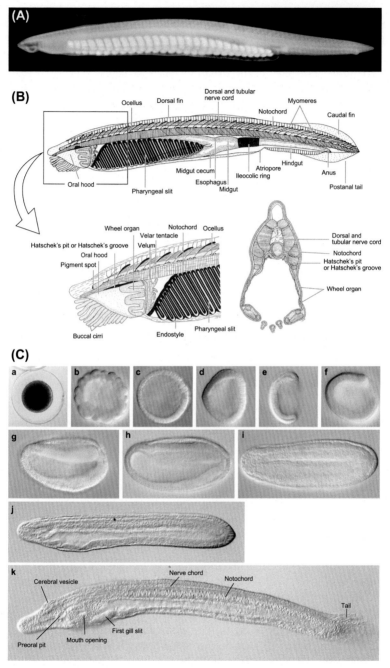

FIGURE 1.4   Cephalochordates. (A) The amphioxus, *Branchiostoma floridae* (male). *(Courtesy of Jr-Kai Yu.)* (B) Internal organs and tissues of amphioxus. An enlargement of structures in the anterior pharyngeal region (lower left) and a cross section of the middle of the body (lower right). *(From Kardong, K.V., 2014. Vertebrates: Comparative Anatomy, Function, Evolution, seventh ed. McGraw-Hill, New York, NY.)* (C) Embryogenesis of *B. floridae*. (a) Fertilized egg, (b) 128-cell embryo, (c) late blastula, (d) early gastrula, (e) mid gastrula, (f) late gastrula, (g) early neurula, (h) mid-neurula, (i) late neurula, (j) early larva (L1 stage), and (k) 36-h larva (L2 stage). *(Courtesy of Jr-Kai Yu.)*

projecting digestive cecum or hepatic cecum arises, which is proposed to represent an evolutionary forerunner of the vertebrate liver and pancreas.

The central nervous system of cephalochordates is very simple. A dorsal nerve cord extends most of the length of the body and is generally expanded slightly to form a cerebral vesicle at the base of the oral hood (Fig. 1.4B). The epidermis is rich in sensory nerve endings, most of which are probably tactile and therefore important in burrowing. Some lancelets have a single, simple eyespot near the anterior end of the dorsal nerve cord. Many adult organs and tissues resemble those of vertebrates, such as the hemal system. The lancelet circulatory system comprises a set of closed blood vessels similar to those of fish, although lancelets do not have hearts.

Early amphioxus development takes a form that is intermediate between those of ambulacrarians and vertebrates, with a sea-star–like blastula, a urochordate-like gastrula, and a vertebrate-like neurula (Fig. 1.4C; Whittaker, 1997; Bertrand and Escriva, 2011). Elongation, somitogenesis, and pharyngogenesis produce a long, thin larva with a body plan similar to that of the adult, although the mouth and first gill slits are on the left and right sides of the pharynx, respectively (Fig. 1.4C). The pelagic larvae almost resemble juveniles. Metamorphosis is driven by thyroid hormone and involves resolution of larval asymmetries and the development of adult structures associated with burrowing, including the atrium and oral tentacles.

1.3.4 Urochordates (Tunicates)

I prefer to use *urochordates* in relation to chordate evolution whereas *tunicates* is preferable to use to explain this animal group identity. All tunicates or urochordates are marine filter feeders with gill slits and an endostyle. Tunicates comprise three classes of approximately 3000 extant species—the Ascidiacea (ascidians; sessile), the Appendicularia (larvaceans; planktonic, tadpole-like juveniles), and the Thaliacea (salps; planktonic, barrel shaped). Among deuterostomes they are unique in various aspects of their biology. First, the entire adult body is invested with a thick covering, the tunic (or test) (Fig. 1.5A and B), from which the name tunicate is derived. A major constituent of the tunic is tunicin, a type of cellulose. Tunicates are the only animals that can independently synthesize cellulose, apparently taking advantage of a horizontal transfer of cellulose synthase genes (Chapter 11). Second, the tunic may function as an outer protective structure, similar to a mollusk shell, and has undoubtedly influenced the development of various lifestyles in this group. They run the gamut of life history strategies, including solitary, colonial, sessile, free-swimming, and asexual-budding forms. As a result, their adult body plans can vary dramatically between classes. Chordate affinities are most evident in the larval form (Satoh, 1994, 2014; Jeffery and Swalla, 1997). An ascidian tadpole has a tubular nerve cord, a notochord, and a postanal tail (Fig. 1.5M and N). These features regress at metamorphosis, leaving the branchial basket, a small nerve ganglion, and

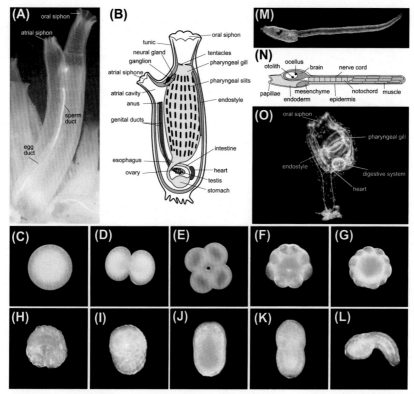

FIGURE 1.5   Urochordates (Tunicates). The ascidian, *Ciona intestinalis*. (A) An adult with oral (incurrent) and atrial (outcurrent) siphons. The *white* duct is the sperm duct and the *orange* duct paralleling it is the oviduct. (B) Diagram illustrating adult organs and tissues. (C–L) Embryogenesis. Embryos were dechorionated to reveal their outer morphology. (C) Fertilized egg, (D) 2-cell embryo, (E) 4-cell embryo, (F) 16-cell embryo, (G) 32-cell embryo, (H) gastrula (~150 cells), (I, J) neurula, (K, L) tailbud embryos, and (M) tadpole larva. (N) Diagram illustrating larval organs and tissues. (O) A juvenile a few days after metamorphosis, with internal structures labeled.

the endostyle as the only chordate characters in the adult. Third, they exhibit very rapid, highly determinate, early development (Fig. 1.5C–L), forming tadpole-like larvae (Fig. 1.5M and N). Tunicate larvae swim without opening their mouths while searching for substrates suitable for attachment where they metamorphose into juveniles (Fig. 1.5O). With its tadpole-like larvae, many basic chordate features, and genomic simplification, the ascidian, *Ciona intestinalis*, is a model organism for exploring molecular mechanisms underlying the origin of chordates and the evolution of vertebrate features (Christiaen et al., 2009; Lemaire, 2011; Satoh, 2014).

An individual ascidian has two openings, an incurrent oral (branchial) siphon and an outcurrent atrial siphon (Fig. 1.5A and B). The mouth behind the oral siphon leads to a large pharynx, or branchial basket—a chamber perforated

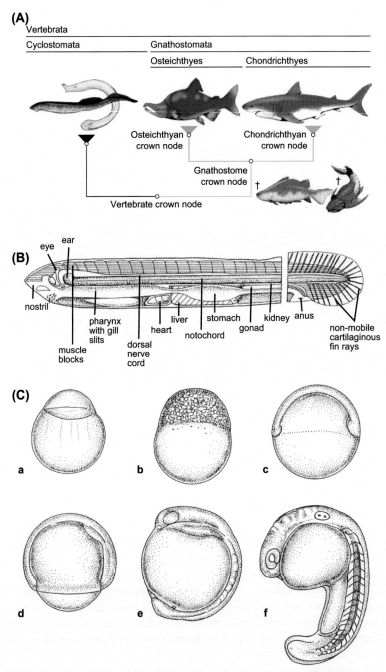

(A)

Vertebrata

Cyclostomata | Gnathostomata

Osteichthyes | Chondrichthyes

Osteichthyan crown node

Chondrichthyan crown node

Gnathostome crown node †

Vertebrate crown node

(B)

eye | ear | nostril | pharynx with gill slits | muscle blocks | dorsal nerve cord | heart | liver | notochord | stomach | gonad | kidney | anus | non-mobile cartilaginous fin rays

(C)

a b c

d e f

FIGURE 1.6 **Vertebrates.** (A) Vertebrates comprise agnathostomes (cyclostomes) and gnathostomes (osteichthyans and chondrichthyans). Crown-, total-, and stem-group concepts provide a useful framework for navigating evolutionary trees that include fossils. Crown groups comprise the last common ancestor of a group of living species, plus all of its descendants, both fossil and modern.

by numerous dorsoventral rows of gill slits called stigmata (Fig. 1.5B). Along the ventral margin of the branchial basket is the endostyle, which secretes large quantities of mucus used for capturing food particles (Fig. 1.5B). The endostyle contains iodine and has an evolutionary relationship with the vertebrate thyroid gland (Chapter 7). The digestive tract leads to a stomach at the bottom of the U-shaped digestive loop, followed by an intestine that terminates at the anus, which opens into the atrial cavity (Fig. 1.5B). The adult nervous system consists of a single cerebral ganglion lying between the two siphons and an adjacent neural gland (Fig. 1.5B). Several nerves extend from the ganglion to various parts of the body, including the muscles, pharynx, viscera, gonads, and siphons. By contrast, the neural gland, which has possible homology with the vertebrate anterior pituitary gland, leads through a duct to the pharynx, just behind the mouth. The open circulatory system is well developed and consists of a short, tubular heart and numerous blood vessels. The posteroventral heart lies near the stomach and behind the pharyngeal basket (Fig. 1.5B and O). The heartbeat and the direction of blood flow reverse periodically, and the circulatory system contains several different types of blood cells or coelomic cells with specialized functions.

1.3.5 Vertebrates

Among metazoans, vertebrates have evolved the most complex traits embryologically, morphologically, and physiologically (eg, Kardong, 2014). Detailed description of them would require another book. They include approximately 64,000 described species—more than half of them bony fishes. The classification of vertebrates has recently been discussed based on various characteristics, including molecular data. Vertebrates are traditionally divided into two groups: agnathans (jawless) and gnathostomes (jawed) (Fig. 1.6A). Agnathans include

◀ The gnathostome crown group includes the last common ancestor of osteichthyans (represented by a salmon) and chondrichthyans (represented by a shark) plus all of its descendants and comprises all of the *green* and *orange* parts of the tree. Here, the gnathostome total group is represented by all *colored* parts of the tree. Stem groups are equal to a clade's total group minus its crown group, shown here by the *pink* lineage connecting the vertebrate and gnathostome crown nodes. Jawed vertebrates include the gnathostome crown and the upper reaches of the gnathostome stem. The lower part of the gnathostome stem is populated by jawless ostracoderms, which are more closely related to jawed vertebrates than to modern jawless fishes. *(From Brazeau, M.D., Friedman, M., 2015. The origin and early phylogenetic history of jawed vertebrates. Nature 520, 490–497.)* (B) The hypothetical "basic" body plan of vertebrates, shown in longitudinal section. *(From Jefferies, R.P.S., 1986. The Ancestry of the Vertebrates. British Museum (Natural History), London, UK.)* (C) An overview of early zebrafish development. (a) The newly fertilized egg consists of large region rich with yolk vesicles and a smaller yolk-free blastodisc. (b) The blastula with an undifferentiated ball of cells. (c) The gastrula after epiboly. Concurrent movements of involution and convergent extension produce different germ layers and the primary embryonic axis. (d–f) A tailbud embryo and early larvae. The notochord differentiates, somites appear sequentially along the axis, and the central nervous system and sensory organs become prominent. *(From Kimmel, C.B., Ballard, W.W., Kimmel, S.R., Ullmann, B., Schilling, T.F., 1995. Stages of embryonic development of the zebrafish. Developmental Dynamics 203, 253–310.)*

hagfish and lampreys. They are fish that lack jaws and rigid, hinged elements supporting the border of the mouth. They are sometimes called cyclostomes and are regarded as ancestral to vertebrates, although the few living representatives are highly specialized for unusual feeding modes. Characters that show similarities between vertebrates and other chordates are found in embryonic stages of various vertebrate groups, as represented by the ammocoete larvae of lampreys.

Unlike agnathans, gnathostomes have jaws and include six classes: Chondrichthyes (cartilaginous fishes such as sharks and rays), Osteichthyes (bony fishes; Fig. 1.6A), Amphibia (frogs, toads, urodeles, and caecilians), Reptilia (crocodilians, turtles, lizards, snakes, and amphisbaenians), Aves (birds), and Mammalia (mammals), although recent studies suggest that the Class Aves actually belongs to the Reptilia (Hedges and Poling, 1999; Alföldi et al., 2011). Strictly speaking, although the term *Vertebrata* is widely used to cover this taxon, cyclostomes do not actually develop vertebrae. Although the term *Craniata* (in contrast to *Acraniata*, or lancelets) covers all three groups, *Vertebrata* is generally preferred.

Unlike other deuterostome groups, vertebrates are uniformly motile, and they are solitary rather than colonial. All vertebrates have body plans characterized by at least a vertebral column; a dorsal, central nerve tube; and gills (Kardong, 2014; Fig. 1.6B). It is well accepted that vertebrates are distinguished from other chordates by the elaboration of the head region with an anterior central nervous system and paired sense organs. Many of these elaborations are linked to the innovation of the neural crest, which first appears in vertebrates. In addition, an endoskeleton, an adaptive immune system, a gene content brought by two rounds of genome-wide gene duplication, and a placode are also vertebrate-specific features, which will be discussed in detail in Chapter 10.

Reflecting various adult vertebrate body plans, the mode of embryogenesis differs widely among classes. In zebrafish, early cleavage takes place in embryonic cells located near the animal pole (Fig. 1.6C). Epibolic movement of cells covers two-thirds of the blastula, which is followed by gastrulation, neurulation, and construction of the tailbud embryo (Langeland and Kimmel, 1997).

1.4 CONCLUSIONS

Chordate origins and evolution have to be discussed in relation to five deuterostome taxa: echinoderms, hemichordates, cephalochordates, tunicates, and vertebrates. The first two are associated with deuterostome evolution and chordate origins, and the last three are associated with chordate origin and evolution. However, as described here, the five classes are distinguished by their own characteristic features. To date, no intermediate taxa exist in the fossil record. This suggests that we need to consider very carefully which features are shared as a result of common ancestry and which lineage-specific features each taxon evolved.

Chapter 2

Hypotheses on Chordate Origins

The appearance of chordates has been debated for more than 150 years, and many hypotheses have been offered to explain this evolutionary event. These include the annelid hypothesis, the auricularia hypothesis, the calcichordate hypothesis, the enteropneust hypothesis, the inversion hypothesis, and the aboral dorsalization hypothesis. Each had a theoretical basis at the time it was proposed.

Although the idea of evolution existed among Greek philosophers, the concept of evolution by means of natural selection originated with Charles Darwin in 1859. His insight into animal evolution included questions regarding the timing of evolutionary events occurred, what caused the changes to occur, and what mechanisms produced the changes. On the basis of this concept, Ernest Haeckel (1866) produced a famous phylogeny entitled the "pedigree of man" (Fig. 2.1). Although in one sense this phylogeny is not correct, because human beings are not the only pinnacle of evolution, the tree of life propounded a practical idea as to how chordates evolved from a common ancestor shared with other metazoans (see the discussion of the history of taxonomic classification and chordate terminology in Section 1.2.3).

Since then, chordate evolution has been investigated and vigorously debated for more than 150 years because the debate is pertinent to the origin of vertebrates, including man. At one time or another, almost all representative nonchordate metazoans including nemerteans, lophophorates, molluscs, annelids, and crustaceans have been discussed in relation to chordate origins (eg, Lovtrup, 1977; Gould, 1977; Willmer, 1990; see detailed discussion in books by Hall, 1999 and Gee, 1996). In Darwin's day, researchers felt that there was a large gap between invertebrates and vertebrates. Alexander Kowalevsky showed otherwise in 1866 and 1867 when he discovered that tunicates and amphioxus are related to vertebrates by virtue of the presence of a notochord and axial structures, including the central nervous system (CNS), in the larvae and/or adults (Kowalevsky, 1866a, 1867). Approximately 150 years later, our understanding of the origin and evolution of chordates has progressed tremendously with data from molecular developmental biology. However, debate still swirls, reflecting the profundity of this biological conundrum.

This chapter discusses several hypotheses proposed to answer questions regarding chordate beginnings. How meaningful are similarities of body plans? Were adult ancestral chordates sessile or free-living? Did chordates originate

Chordate Origins and Evolution. http://dx.doi.org/10.1016/B978-0-12-802996-1.00002-X

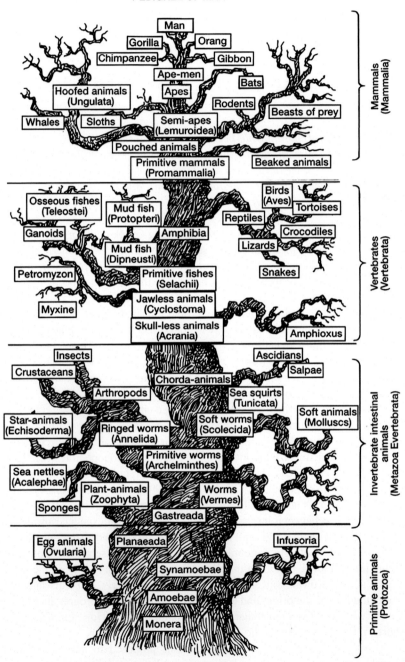

FIGURE 2.1 Haeckel's phylogeny and tree of life. In this tree, "Pedigree of Man," Haeckel espoused the view that humans represent the culmination of evolution. Here the Acrania is basal to vertebrates. The Tunicata is closest to vertebrates among chorda-animals, and soft worms are close to chorda-animals. *(From Haeckel, E., 1866. Generelle Morphologie der Organismen, Verlag von Georg Reimer, Berlin, Germany.)*

through an inversion of the dorsoventral (DV) axis of nonchordate animals? How did the chordate body plan, especially its adult form, originate from the common ancestor(s) of deuterostomes? I will discuss these issues in light of findings in embryology or evolutionary developmental biology. The proposed hypotheses are not always independent, and supporting arguments for them frequently overlap.

2.1 THE ANNELID THEORY

By carefully comparing the anatomy of annelids and arthropods with that of vertebrates, Geoffroy St-Hilaire (1822) noticed that the basic body plans were similar, but that the position of the nerve cord was different (Fig. 2.2A). Both annelids and vertebrates share segmented body plans with an anteroposterior (AP) axis and a well-developed, anterior CNS. However, in annelids the nerve cord runs ventral to the digestive tract; therefore these groups are sometimes called Gastroneuralia. In contrast, in vertebrates the nerve cord runs dorsal to the digestive system; hence, they are sometimes called Notoneuralia. The annelid theory had its roots in protostome anatomy. Anton Dohrn (1875) originally proposed the inversion of the body plan, suggesting that during evolution of vertebrates from annelids, the mouth (old) was first positioned on top of the head (Fig. 2.2B(a)) and, necessitating the formation of a new mouth on the ventral side of the body (Fig. 2.2B(b)), the old mouth disappeared whereas the new one persisted (Fig. 2.2B(c)). That is, a worm inverted the DV axis of the body while evolving into a vertebrate. This theory was discussed further by Gaskell (1890) in relation to crustacean anatomy.

Researchers had forgotten the annelid theory for long time. However, as discussed in Section 2.5, with the support of molecular data, this theory has been revitalized as "the new inversion theory (hypothesis)."

2.2 THE AURICULARIA HYPOTHESIS

Around the turn of the 20th century, echinoderms and hemichordates were gradually recognized as deuterostomes, presumably sharing their ancestry with chordates. It seems natural that researchers sought the origin of chordates among their cousins, echinoderms and hemichordates. Garstang (1928a, b) proposed one hypothesis based on this view. As discussed in Section 1.3.1, echinoderms have a spectacular variety of larval forms, from the ophiopluteus and echinopluteus, which share similar, elaborate skeletons, to the asteroid bipinnarian and holothuroid auricularian larvae. However, despite the morphological and developmental dissimilarities among echinoderm larvae, they share similar convoluted ciliary bands for swimming and feeding, suggesting the existence of an ancestral "dipleurula" (small two-sided) larval form, from which ambulacrarian larval diversity arose. In addition, the organization of the hemichordate tornaria larval body plan is very similar to that of echinoderm larvae (Section 1.3.2), although the tornaria has a robust additional posterior band of compound cilia.

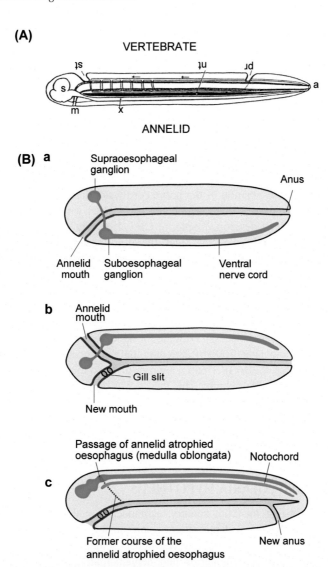

FIGURE 2.2   The essence of the annelid hypothesis. (A) Romer's description (1970) of the annelid hypothesis to illustrate the supposed transformation of an annelid worm into a vertebrate. In a normal position, this represents the annelid with a "brain" (s) at the anterior end and a nerve cord (x) running along the underside of the body. The mouth (m) is on the underside of the animal, the anus (a) at the end of the tail; the blood flows anteriorly on the dorsal side of the body, posteriorly on the venter (indicated by *arrows*). Turn the book upside down—and now we have a vertebrate, with the nerve cord and blood flow reversed. However, it is necessary to build a new mouth (st) and anus (pr) close to the old ones. Annelids really have no notochord (nt), and the supposed change is not as simple as it seems *(From various sources)*. (B) An annelid with a central nervous system (CNS; *green*) comprising supraoesophageal and suboesophageal ganglia, circumoesophageal axonal projections, and a ventral nerve cord. (a, b) Dorsoventral inversion produces a new foregut (purple) penetrated by gill slits. (b, c) The annelid-to-vertebrate transition. The new foregut persists, but the old one atrophies, permitting union of the supra- and suboesophageal ganglia into a vertebrate-like brain. A notochord (*blue*) originates from connective tissue surrounding the nerve cord, and a new anus opens. *(From Holland, N.D., Holland, L.Z., Holland, P.W., 2015. Scenarios for the making of vertebrates. Nature 520, 450–455.)*

Walter Garstang (1928a,b) proposed the auricularia hypothesis in an attempt to explain how chordate body plans originated from a deuterostome common ancestor by emphasizing the significance of changes in larval forms (Fig. 2.3A). The dipleurula larva has bilateral symmetry and a one-way gut. Near the mouth is an aboral band of feeding cilia and across its lateral surface a long, wavy row of circumoral cilia (Fig. 2.3A). According to this view, ptero-branch-like, sessile animals with auricularia-like larvae led to primitive ascid-ians (as the most recent common ancestor of chordates, at that time) through morphological changes in both larvae and adults (Fig. 2.3A). Therefore this hypothesis falls under the sessile ancestor scenario discussed in Section 2.4. According to Garstang's view, adults changed their feeding apparatus from external tentacles to internal branchial sacs. In larvae, the ancestor's circum-oral, ciliated bands and their associated underlying nerve tracts moved dorsally to meet and fuse at the dorsal midline, forming a dorsal nerve cord in the chor-date body (Fig. 2.3A). At the same time, the aboral ciliated band gave rise to the endostyle and ciliated tracts within the pharynx of the chordate (Fig. 2.3A). Claus Nielsen (1999) proposed a revised version of this hypothesis in which the chordate CNS evolved from the postoral loop of the ciliary band in a dipleurula larva (Fig. 2.3B). During these transitions, the larval body elongated and devel-oped muscle and a notochord.

2.3 THE CALCICHORDATE HYPOTHESIS

The calcichordate theory was proposed rather recently by paleontologist Richard Jefferies (1986). This hypothesis holds that chordates evolved from the lineage of an echinoderm fossil named *Mitrata* [see detailed discussions of Jefferies et al. (1996) and Gee (1996)]. The Paleozoic echinoderm-like fossil carpoids, *Cornuta* and *Mitrata*, are grouped together in a clade called Calcichordata (classically known as stylophorans; Fig. 2.4). According to this hypothesis, calcichordates are a paraphyletic array of stem chordates, stem cephalochordates, stem tunicates, and stem vertebrates, the calcitic skeleton of which has been lost several times (for the stem group concept, see Section 3.2 and the legend of Fig. 1.6). The mitrates and chordates are called Dexiothetes, dexiothetism being a synapomorphy for the clade (Fig. 2.4B). The appendage of the carpoids is regarded as a tail, with the central canal probably containing a notochord (Fig. 2.4C). The large orifice is likely a mouth with many of the slits along the side assumed to be gill slits. Ancestors of chordates underwent a profound remodeling of their body plan, becoming dexiothetetic and losing their calcitic skeletons. Because the Cornuta were interpreted as lying with the flat side ventrally, Jefferies suggested that in the Mitrata, the flat side was dor-sal and the convex side ventral whereas the tail was curved underneath to pro-vide forward thrust (Jefferies, 1986, 1991). Many mitrates are preserved with the tail underneath. With the advent of molecular data, this hypothesis has lost credibility. For example, comparative genomics indicate that echinoderms are

(A)

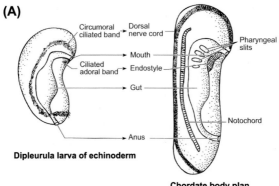

Dipleurula larva of echinoderm

Chordate body plan

(B) ENTEROPNEUSTA   CHORDATA

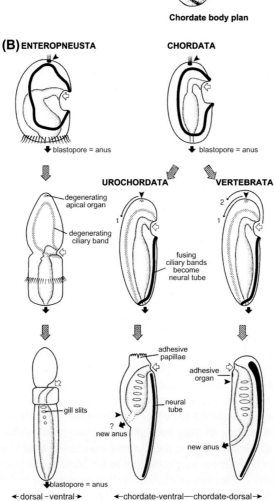

only a deuterostome group in which a calcareous endoskeleton evolved from the beginning (Chapter 5).

Although the calcichordate hypothesis once provoked heated controversies, many recent studies do not support it. However, even if it proves erroneous, the Calcichordata theory illustrated two significant points in relation to the origin and evolution of chordates. First, extinct echinoderms had a pharynx; and second, the close relationship between tunicates and vertebrates resulted in a group called Olfactores because some stylophorans that were thought to be stem tunicates display internal structures that resemble vertebrate olfactory organs.

2.4 THE PEDOMORPHOSIS SCENARIO: WAS THE ANCESTOR SESSILE OR FREE-LIVING?

To explain the auricularia hypothesis, Garstang (1928a,b) hypothesized that the chordate ancestor was sessile, similar to extant ascidians, and that two rounds of pedomorphosis occurred before the advent of motile lancelet-like ancestors. Since then, the question of whether the chordate ancestor was sessile or free-living has been extensively discussed (see discussion by Lacalli, 2005; Swalla, 2006; Fig. 2.5). As described in Section 1.3.2, extant hemichordates consist of two groups with different lifestyles: the sessile, colonial pterobranchs and the free-living enteropneusts (acorn worms). In addition, chordates include sessile tunicates and free-living, solitary cephalochordates. Therefore various authors have addressed the question of whether the chordate ancestor(s) was sessile or free-living. According to one scenario, supported by Garstang (1928a,b), Berrill (1955), and Romer (1967), and therefore tentatively dubbed the GBR hypothesis, ancestral deuterostomes were sedentary, tentaculate animals with pelagic larvae, similar to modern pterobranchs, which evolved into sedentary ascidians (Fig. 2.5A). The motile, free-living lifestyle of cephalochordates and vertebrates was believed to have evolved twice from

FIGURE 2.3 The auricularian hypothesis and related hypotheses. (A) Diagram showing Garstang's theory. The proposed common ancestor of chordates (left) was bilaterally symmetrical and had the external appearance of a young echinoderm larva. The ancestor's circumoral ciliated bands and their associated underlying nerve tracts moved dorsally to meet and fuse at the dorsal midline, forming a dorsal nerve cord in the cephalochordate body plan. The aboral ciliated band gave rise to the endostyle and ciliated tracts within the pharynx of the chordates. Garstang noted that the appearance of pharyngeal slits improved efficiency by providing a one-way flow for the food-bearing stream of water. A notochord appeared later, and with swimming musculature it is advantageous for locomotion in the larger organism *(From various sources)*. (B) Nielsen's schematic representation of the transformation of ambulacrarian larvae to chordate body plans. Urochordates are represented by a tadpole-like larva. *Arrows 1* and *2* denote movements of the apical pole and the mouth, respectively. The apical pole is indicated by *arrowheads*, the mouth by *white arrows*, and the anus by *black arrows*. The intraepithelial nervous system of the adult enteropneust has been omitted. The position of the new anus in the ancestral urochordate larva is uncertain because it is unknown when the atrium evolved. *(From Nielsen, C., 1999. Origin of the chordate central nervous system – and the origin of chordates. Development Genes and Evolution 209, 198–205.)*

(A)

(B)

(C)

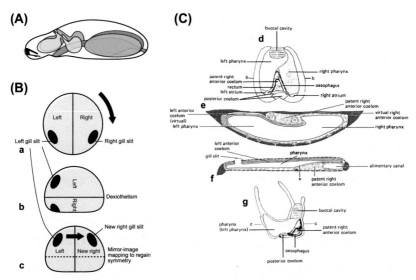

FIGURE 2.4 Calcichordate hypothesis. (A) A hypothetical larva of primitive echinoderms. (B) Diagram showing changes in the left-right asymmetry in Dexiothetes (echinoderms + chordates) results from dexiothetism—an episode in their ancestry when an animal resembling the recent pterobranch, *Cephalodiscus*, lay right-side downward on the sea floor. (c) Mitrate left-right organ paring. (C) The body plan of the carpoids. (d) Head of the mitrate (dorsal view). (e) Transverse section through b–b in d. (f) Transverse section of the cornuted through c–c in g. (g) Head of the cornuted (dorsal view). In the origin of mitrates, a right pharynx branched from the left pharynx, lifted up the cavity and contents of the right mandibular somite, and faced them into a medial position hanging from the top of the head. *Black* indicates skeleton. (*From Jefferies, R.P.S., Brown, N.A., Daley, P.E.J., 1996. The early phylogeny of chordates and echinoderms and the origin of chordate left-right asymmetry and bilateral symmetry. Acta Zoologica (Stockholm) 77, 101–122.*)

a motile larval stage of the sedentary, tentaculate ancestor by pedomorphosis: first, at the larval stage of pterobranchs and second at tadpole larval stage of ascidians (Fig. 2.5A). Pedomorphism is a form of heterochrony roughly equivalent to neoteny, in which the larval stage becomes sexually mature and replaces the adult. The latter looks like tunicate larvaceans, in which adult organs develop in the trunk region of tadpole-like juveniles, and this may be a good example of a pedomorphogenetic transition.

Other researchers including Bone (1960), Tokioka (1971), Jollie (1973), Salvini-Plawen (1999), and Cameron (2005) proposed an alternative scenario, involving the progressive evolution of motile adults (Fig. 2.5B). According to this scenario, the chordate ancestor was free-living and vermiform, and the sequence of ancestral forms is thought to have consisted of motile, bilaterally symmetrical organisms, as opposed to larvae (Brown et al., 2008; Satoh, 2008). Motile forms such as enteropneust hemichordates and cephalochordates are typically considered close to the main lineage whereas urochordates are viewed as more distant.

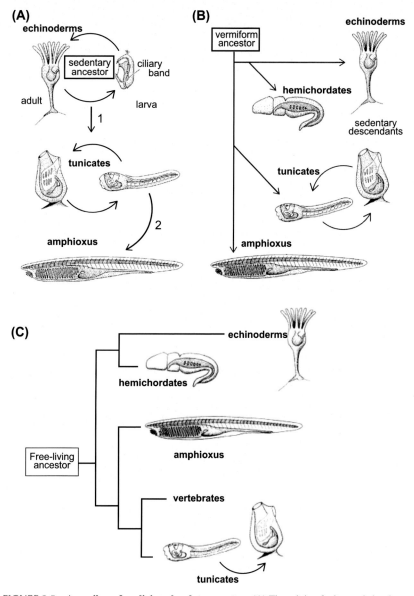

FIGURE 2.5   A sessile or free-living chordate ancestor. (A) The origin of advanced chordates, represented here by amphioxus, from a series of sessile ancestors, which is essentially the GBR (Garstang-Berrill-Romer) hypothesis, as expounded variously by Garstang, Berrill, and Romer. Ancestral adults remain sessile whereas their motile larvae evolve progressively in two steps. The first is to (1) a tadpole-like protochordate and then to (2) a more advanced and fully motile form. *(From Lacalli, T.C., 2005. Protochordate body plan and the evolutionary role of larvae: old controversies resolved? Canadian Journal of Zoology 83, 216–224.)* (B) The main alternative to the GBR hypothesis. Throughout the evolutionary sequence, the ancestral adult is assumed to be a motile, somewhat vermiform animal, either swimming or burrowing. A selection of its descendants

are shown, but the ancestral form is not because its exact nature remains a matter of conjecture. Ancestral larvae, assuming they existed, can be disregarded in scenarios of this type, and all sedentary forms are then derived. In the case of tunicates, the tadpole-like larva would be an evolutionary relic, probably much reduced from its original form, rather than a progenitor of segmented chordates. *(From Lacalli, T.C., 2005. Protochordate body plan and the evolutionary role of larvae: old controversies resolved? Canadian Journal of Zoology 83, 216–224.)* (C) A novel and likely scenario for the origin and evolution of chordates. The deuterostome ancestor was a motile, free-living ancestor. The motile enteropneust-like ancestor gave rise to cephalochordates, and then to vertebrates in a direct lineage, whereas urochordates were derived from a sessile adult form. *(From Satoh, N., 2008. An aboral-dorsalization hypothesis for chordate origin. Genesis 46, 614–622.)*

Historically the first scenario of sessile ancestry received much support because ascidians had long been accepted as the most basal chordates. The "tadpole larva" was frequently cited in this regard. Despite the long-recognized unity of chordates, our understanding of the relationship among the three chordate taxa has undergone a dramatic shift in the past 10 years because of the emergence of deep genomic datasets. Molecular phylogenetics and genomics have rapidly convinced most researchers that cephalochordates, represented by amphioxus, diverged first (Chapter 4), supporting a free-living ancestor of chordates (Fig. 2.5C). In this context, references to "tadpole-like larvae" should be changed to "fish-like larvae," or else these names should be used in different contexts, the former in association with olfactory evolution and the latter with regard to chordate origin.

2.5 THE NEW INVERSION HYPOTHESIS

In the 1990s, advances in developmental genetics set the stage for a revival of the annelid theory, namely the discovery of genes responsible for DV axis formation (Arendt and Nübler-Jung, 1994; Holley et al., 1995; De Robertis and Sasai, 1996). The frog *bmp4* gene is expressed ventrally and found to have ventralizing activity whereas its fly homolog, the *dpp* gene, is expressed dorsally and has dorsalizing activity (see Fig. 8.3). In addition, the frog *chordin* gene is expressed dorsally and has dorsalizing activity whereas its fly homolog, the *sog* gene, is expressed ventrally and has ventralizing activity. Therefore *bmp4/dpp* and *chordin/sog* constitute homologous pairs of antagonists that apparently promoted reversal of the DV axis between flies and frogs (see Fig. 8.3). Arendt and Nübler-Jung (1994) interpreted this pattern as support for homology between arthropod and vertebrate nerve cords and indicative of a DV inversion of the body during the invertebrate-to-vertebrate transition. Additional support came from the finding that neural progenitor cells in the CNS were organized in longitudinal bands, each characterized by a distinctive suite of gene expression that was homologous between flies and vertebrates, and that gene expression in these bands was mediolateral in both organisms (Arendt and Nübler-Jung, 1999). This and the enteropneust hypothesis that follows are the main discussion points of this book and will be addressed in Chapters 8 and 9.

2.6 THE ENTEROPNEUST HYPOTHESIS

The enteropneust hemichordate body is composed of three parts: a flattened oral shield corresponding to the enteropneust proboscis, a collar extending into tentacle-fringed arms, and a trunk (Fig. 1.3B). Historically, before Garstang's auricularian hypothesis, William Bateson (1886) and Alexander Kowalevsky (1886a) noted many similarities between enteropneusts and chordates: the stomochord corresponded to a vertebrate notochord, the collar cord (which Bateson considered dorsal) corresponded to the vertebrate CNS, and the pharyngeal gill slits in both groups were homologous. In addition, Thomas H. Morgan (1891) pointed out the occurrence of vertebrate-like neurulation in hemichordate adults. As such, an enteropneust was much like a vertebrate (eg, Ruppert, 2005), except that it lacked segmented musculature along the anterior–posterior axis. Accordingly, Bateson coined the name "hemi-(half)-chordates" (1886). According to the original enteropneust theory of Bateson, the body axis of enteropneusts was not inverted relative to that of vertebrates.

2.6.1 A New Enteropneust Hypothesis

In relation to the inversion hypothesis, an overarching question was when and where in metazoan phylogeny the DV axis inversion occurred. Was it between the split of protostomes and deuterostomes or somewhere in the deuterostome lineage? Although it will be discussed in detail in Chapter 8, recent gene expression studies suggest that the inversion appears to have occurred at the time of the separation of ambulacrarians and chordates. In echinoderms and hemichordates, *BMP* is expressed on the aboral side of the embryo and *chordin* on the oral side (Duboc et al., 2004; Lowe et al., 2006). In contrast, in cephalochordate embryos, *BMP* is expressed on the ventral side and chordin on the dorsal side, as in vertebrates (Yu et al., 2007). *Saccoglossus kowalevskii* is an acorn worm in which fertilized eggs develop directly into adults without a tornaria larval stage. In this species, the oral-aboral orientation of the embryo becomes a ventral-dorsal orientation in the adult. Therefore DV axis inversion appears to have occurred during the evolution of chordates.

After the finding of DV axis inversion timing, Bateson's hypothesis was revitalized by Nübler-Jung and Arendt (1996) (Fig. 2.6). They proposed that enteropneusts had an annelid-like CNS comprising three contiguous nerve tracts (the collar cord, the circumenteric nerve ring, and the trunk ventral nerve cord), all recognizable by their giant nerve fibers. Thus enteropneusts appeared to support a revival of the earlier annelid theory by approximating an intermediate stage in the conversion of a complex bilaterian (sometimes called urbilaterian) into a vertebrate (Fig. 2.6B). Because the CNS is oriented as in annelids, the conversion into a vertebrate-like descendant would require DV inversion, conflicting with Bateson's original scenario (Fig. 2.6). This hypothesis will also be discussed in Chapters 8 and 9.

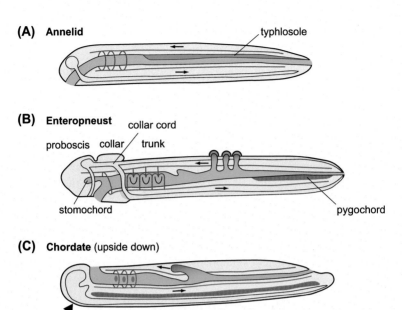

FIGURE 2.6 A modern inversion hypothesis. Comparison of generalized body plans of an annelid (A), an enteropneust (B), and a chordate (C) with the chordate turned upside down. The direction of blood flow (*arrows*) and the relative positions of the gut (*green*), the principal nerve cord (*yellow*), and the chordate notochord/enteropneust pygochord (*brown*) are the same. Likewise, the paired hepatic outpockets in the enteropneust lie in a region comparable to the typhlosole (*blue*) in the annelid and to the liver in the chordate (*dark green*). Note that the proposed homology between the enteropneust stomochord and the chordate notochord has been repeatedly contested. The *arrowhead* indicates the position of the mouth in the putative gastroneuralian ancestor of chordates. Note that the functional mouth of the chordate is a new mouth. (*Modified from Nübler-Jung, K., Arendt, D., 1996. Enteropneusts and chordate evolution. Current Biology 6, 352–353.*)

2.7 THE ABORAL-DORSALIZATION HYPOTHESIS

I proposed the aboral-dorsalization hypothesis in 2008 (Satoh, 2008). In contrast to most of the hypotheses described earlier herein, which rely upon the comparison of adult morphology, the aboral-dorsalization hypothesis relies on embryology, or the mode of early development (Fig. 2.7). I have been working on the gastrula, larva-I, and larva-II developmental biology of ascidians, especially the molecular mechanisms involved in notochord formation (Satoh, 2003, 2014). During my 20 years of research on the chordate notochord, I have come to feel strongly that even if DV axis inversion occurred at a very early phase of chordate evolution, the notochord did not develop just as a result of DV axis inversion. An organ homologous to the notochord had to exist before the inversion. In general, such organs are not found in nonchordate invertebrates, except for possible homology of the stomochord and pygochord in hemichordates and the axochord in annelids (Chapter 9). The pygochord is a rod-like thickening in the posthepatic region of the trunk, mid-ventrally between the intestine and

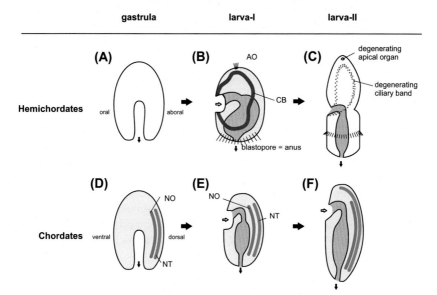

FIGURE 2.7 **Schematic outlining the aboral dorsalization hypothesis.** It explains the development of fish-like larvae with a dorsal neural tube and notochord during chordate evolution. Embryogenesis of (A–C) hemichordates and (D–F) chordates. It should be emphasized that the neural tube and notochord formed on the dorsal side of the embryo are completely novel structures without any relationship to the collar cord and stomochord of enteropneust hemichordate juveniles, respectively. *AO*, apical sensory organ; *CB*, ciliary band; *NT*, neural tube; *NO*, notochord. *White arrows* indicate mouth and *black arrows* anus. See the main text for details. (*From Satoh, N., 2008. An aboral-dorsalization hypothesis for chordate origin. Genesis 46, 614–622.*)

body wall of enteropneusts. The notochord should be considered an evolutionary novelty, found only in chordates. If so, then how did the notochord occur? The notochord provides structural support for the tail-beating locomotion system of fish/tadpole-like larvae. The style of this locomotion is unique to chordates. The essence of the aboral-dorsalization hypothesis is the modification of larval form, from cilia-based locomotion to tail-muscle-based locomotion.

The original aboral-dorsalization hypothesis emphasized the location in which the novel structures are formed (Fig. 2.7). Bilaterian bodies are characterized by three major axes: the AP axis, the DV axis, and the left-right (LR) axis. When the first two axes are established, the third, or the LR axis, is automatically fixed. Early ambulacrarian embryos do not have organs and/or tissues that clearly define the dorsal side. Viewed from the vegetal pole, the early ambulacrarian embryo is radially symmetrical, suggesting that dorsal-midline structures could be formed anywhere. However, because the mouth does not form at the apical site, but rather on one side of the embryo, the mouth-opening side defines the oral side and its opposing, aboral side. For this reason, ambulacrarian embryologists prefer to use "the oral–aboral axis" rather than "the DV axis." (eg, Davidson, 2006). It may be imagined that during embryogenesis of ancestral chordates, the neural tube and notochord could have been formed on any side of the embryo—anterior, posterior, left, right, oral, or aboral. Again, viewing the mode of ambulacrarian gastrulation, the comparatively large archenteron is formed along the AP axis. Therefore formation of the neural tube and notochord in the anterior region of the embryo would hinder proper formation of the mouth whereas in the posterior region they would interfere with the formation of the anus. The neural tube and notochord could form on either the left or right side. However, formation on the left or right side would make the body plan asymmetric along the AP axis, which would not generally be advantageous for strong forward movement. The last possible position is either ventral or dorsal. This choice may be determined by the space available within the embryo. In ambulacrarian embryos, the anterior edge of the archenteron bends ventrally and then connects with the newly formed mouth invagination or stomodeum to form a complete digestive tract. It is natural to wonder if this mode of gut formation might have been retained in embryos of chordate ancestor(s). This means that the venter of the embryo would have had limited available space, whereas the dorsal side of the embryo would have provided more space to create novel structures. Accordingly, the notochord and neural tube formed in the dorsal side of the ancestral embryo. That is, the aboral side of nonchordate deuterostome embryos became the dorsal side of chordate embryos to create fish-like larvae. Thus creation of the notochord and nervous system on the dorsal side of ancestral chordate embryos might have resulted from spatial restrictions. This line of thought will be discussed further in Chapter 8 in relation to embryonic axis formation in deuterostomes.

2.8 CONCLUSIONS

Reflecting morphological, embryological, and evolutionary developmental biological research, many hypotheses have been proposed to explain how chordates originated from extant metazoan taxa. Although the inversion hypothesis is now widely accepted, further discussion is required because these hypotheses are still hypothetical.

Chapter 3

Fossil Records

The emergence and evolution of metazoans may have been found in the fossil record. If so, we may be able to follow changes in deuterostome body plans that gave rise to chordate morphology. In the last decade, using more sophisticated technologies, the discovery of fossilized soft-body parts has significantly advanced our understanding of metazoan paleontology. What insights do these discoveries offer regarding chordate origins and evolution?

It seems natural to think that the history of metazoan evolution should be carefully documented in the fossil record and that we might be able to follow exact changes of metazoan body plans by comparing fossils, but because fossilization is an unlikely event, the fossil record is woefully incomplete. Charles Darwin (1859) noted that metazoan fossils appeared abruptly during the Cambrian Period, but many gaps in the fossil record may eventually be resolved by discoveries of new fossils and better interpretation of them. In the last decade, metazoan paleontology has advanced dramatically with the discovery of soft-bodied fossils from the Burgess Shale of British Columbia, Canada and the Chengjiang fauna of southern China. These advances have been discussed in recent reviews (Clarkson, 1998; Conway–Morris, 2003, 2006; Bengtson, 2005; Benton, 2005; Davidson and Erwin, 2006; Erwin, 2011; Erwin et al., 2011; Janvier, 2015; Brazeau and Friedman, 2015). Using the concept of molecular clocks, a precise temporal framework of the metazoan fossil record, supplemented with geochemical evidence of environmental change, allows us to understand when and how metazoans appeared on the Earth (eg, Peterson et al., 2008). This considerably illuminates patterns of metazoan evolution. However, deuterostome fossils are still scarce, and we await further finds.

3.1 THE CAMBRIAN AND EDIACARAN PERIODS

Biogeologically the history of the Earth is divided into two phases: the Proterozoic [before 542 million years ago (MYA)] and Phanerozoic (after 541 MYA; Fig. 3.1). The former includes the Cryogenian period (before 636 MYA) and Ediacaran period (635–541 MYA). The latter comprises the Cambrian (541–485 MYA), Ordovician (485–443 MYA), Silurian (443–419 MYA), Devonian (419–358 MYA), Carboniferous (358–298 MYA), and Permian Periods (298–252 MYA). Diverse bilaterian clades apparently emerged within the first few million years of the early Cambrian Period (Figs. 3.1 and 3.2). There are distinct,

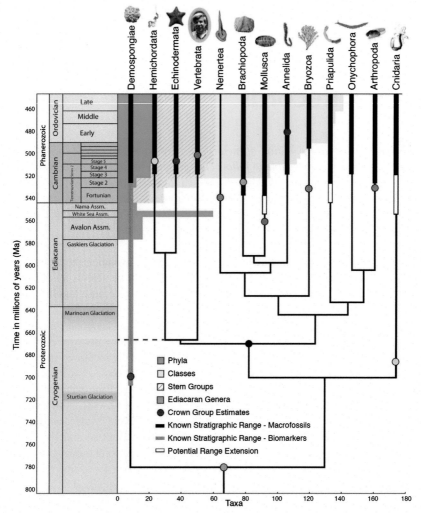

FIGURE 3.1 The fossil record for 13 different animal lineages. Detailed depiction of the fossil record compared to molecular divergence estimates. The dramatic increase in the number of animal fossils (see scale on the bottom) in the Cambrian is shown at the phylum level (*blue*) and at the class level (*yellow*). In this analysis, the origin of deuterostomes was estimated to be approximately 670 million years ago (*dotted red line*). Hatching indicates "stem" lineages that belong to specific phyla but not to any of their extant classes. Shown in *green* is the record of macroscopic Ediacaran fossils. *Thick black lines* represent the fossil records of these 13 lineages through the Cryogenian-Ordovician. Most lineages make their first appearances in the Cambrian, consistent with the known fossil record of all animals (*yellow* and *blue*). Furthermore, the extent of these stratigraphic ranges closely mirrors molecular estimates for the age of each of the respective crown groups (*colored circles*; see also Fig. 3.2), highlighting the general accuracy of the molecular clock. Only cnidarians have an unexpectedly deep crown-group origin, as estimated by the molecular clock. The deep demosponge divergence is apparent from taxon-specific biomarkers (*gray bar*). (*Modified from Erwin, D.H., Laflamme, M., Tweedt, S.M., Sperling, E.A., Pisani, D., Peterson, K.J., 2011. The Cambrian conundrum: early divergence and later ecological success in the early history of animals. Science 334, 1091–1097.*)

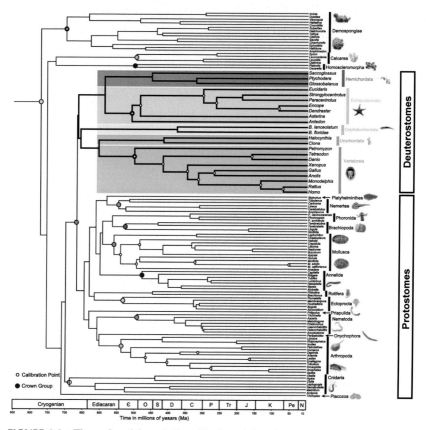

FIGURE 3.2   The early origin and diversification of deuterostomes as inferred from geologic and genetic data. Deuterostome taxa are highlighted as follows: *light green*, echinoderms; *magenta*, hemichordates; *yellow*, cephalochordates; *dark green*, urochordates (tunicates); *blue*, vertebrates. Overlying the geologic record is the pattern of animal origins inferred from the molecular clock. Seven different housekeeping genes from 118 taxa were used to generate this chronogram. Twenty-four calibrations (*open circles*) were treated as soft bounds. This analysis suggests the diversification of cephalochordates was earliest among deuterostomes, ~670 million years ago. All estimates appear to be robust to numerous experimental manipulations performed to assess whether the results were dependent on the parameters used in the analyses. The color of each circle corresponds to the taxonomic bar and label on the far right, with the first appearance of most animal groups at the Ediacaran-Cambrian boundary. Geological period abbreviations: Є, Cambrian; *O*, Ordovician; *S*, Silurian; *D*, Devonian; *C*, Carboniferous; *P*, Permian; *Tr*, Triassic; *J*, Jurassic; *K*, Cretaceous; *Pe*, Paleogene; *N*, Neogene. *(Modified from Erwin, D.H., Laflamme, M., Tweedt, S.M., Sperling, E.A., Pisani, D., Peterson, K.J., 2011. The Cambrian conundrum: early divergence and later ecological success in the early history of animals. Science 334, 1091–1097.)*

large diploblast-like Ediacaran fauna (635–541 MYA; Fig. 3.1). However, Ediacaran originations appear to be relatively few in number, reflecting uncertainties about the phylogenetic placement of most Ediacaran fossils (Clarkson, 1998). In addition, recent findings of embryo-like fossils from Ediacaran fauna have provoked discussions regarding the emergence of metazoans on the Earth (eg, Chen et al., 2002b, 2014). Most Ediacaran metazoans had hard body parts composed of calcium phosphates whereas Cambrian fossils mainly comprised calcium carbonates, presumably because calcium phosphate was more abundant during the Ediacaran Period and was replaced by calcium carbonate during the Cambrian Period. However, we will have to await future developments in Ediacaran fossil research to link the Ediacaran and Cambrian faunas.

According to Erwin et al. (2011), the earliest Cambrian records are fossil traces from bilaterians that had the ability to make complex burrows (540 MYA; Clarkson, 1998; Erwin, 2011; Erwin et al., 2011). A major pulse of metazoan appearances is evident during Cambrian Period Stage 1 (Fortunian; 520 MYA; Fig. 3.1), although the origins of many bilaterian clades are unclear (Davidson and Erwin, 2006). These are mostly known as the "small shelly fossils," leading to extant molluscs, brachiopods, and arthropods. This was followed by the first appearances of many clades during Cambrian Stage 3, documented in the exquisite soft-bodied fossils of the Chengjiang biota. A later, smaller pulse is evident in the Burgess Shale fauna. Most metazoan phyla were evident until the mid-Cambrian Period (Fig. 3.1). New classes continue to arise, albeit at a lower frequency, later into the Cambrian and during the Ordovician radiation (Fig. 3.1). Although extinction has been incessant at lower taxonomic levels, genomic comparisons among surviving members of higher taxa suggest that many of the developmental systems that pattern their body plans have been conserved since early in their history (Davidson and Erwin, 2006; Erwin et al., 2011). In addition to simply describing the fossils, new technical innovations, including high-resolution X-ray microtomography, may reveal more about fossil internal structure and morphology (Dupret et al., 2014).

Therefore diverse bilaterian clades emerged apparently within a few million years during the early Cambrian. Various environmental, developmental, and ecological causes have been proposed to explain this abrupt appearance. One analysis (Fig. 3.2), which combines fossil patterns, molecular diversification, comparative developmental data, and information on ecological feeding strategies, suggests that the major animal clades diverged many tens of millions of years before their first appearance in the fossil record. It is likely that there was a macroevolutionary lag between the establishment of metazoan developmental toolkits (transcription factors, signal pathway molecules, and their regulatory networks) during the Cryogenian (850–635 MYA) and their later ecological success during the Ediacaran (635–541 MYA) and Cambrian (541–488 MYA) Periods. This problem will also be discussed in Chapter 4.

3.2 CROWN, STEM, AND TOTAL GROUPS

In phylogenetic analyses the concepts of crown, stem, and total groups provide a useful framework for navigating evolutionary trees that include fossils (Fig. 3.3; see also Fig. 1.6A). A crown group is the smallest clade that includes all living members of a group and any fossils nested within it. A stem group is a

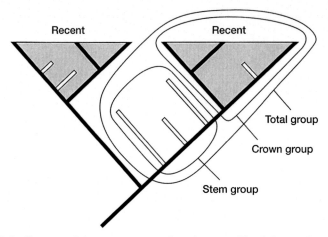

FIGURE 3.3 **Concepts of the stem, crown, and total groups.** The cladogram shows the relationship between extant (*black lines*) and extinct (*white lines*) groups. Stem groups include all intermediate, but now extinct, fossil groups. Together, crown plus stem groups constitute total groups (monophyletic clades). *(From various sources.)*

set of extinct taxa that are not in the crown group but are more closely related to the crown group than to any other. Together crown and stem groups constitute a total group. For example, in Fig. 1.6A, the gnathostome crown group includes the most recent common ancestor of osteichthyans (represented by a salmon) and chondrichthyans (represented by a shark) plus all of their descendants. This concept is useful to infer phylogenetic relationships, vertebrate clade-level evolution with fossil records, or calcichordate theory, discussed in Section 2.3.

3.3 FOSSIL RECORDS OF INVERTEBRATE DEUTEROSTOMES

Echinoderms and vertebrates support their bodies with mineralized exoskeletons and endoskeletons, respectively; therefore they have both left many high-quality fossils. *Pikaia* is a well-known fossil from the Burgess Shale, representing early traces of cephalochordates (Fig. 3.4C). Recently well-conserved tunicate fossils appeared in the Chengjiang fauna. In addition, soft-bodied hemichordate fossils were discovered. Paleontological evidence supports the emergence of all five deuterostome taxa during the early to middle Cambrian Period, although the evolution of early to modern vertebrates appears to have taken a bit longer. The following are brief descriptions of deuterostome fossils in relation to their evolutionary history.

3.3.1 Echinoderms

Echinoderms have one of the best fossil records, which has enabled detailed characterizations of their internal and external morphology (Figs. 3.4A and 3.5; Clarkson, 1998; Smith, 2005; Bottjer et al., 2006). The current view of this

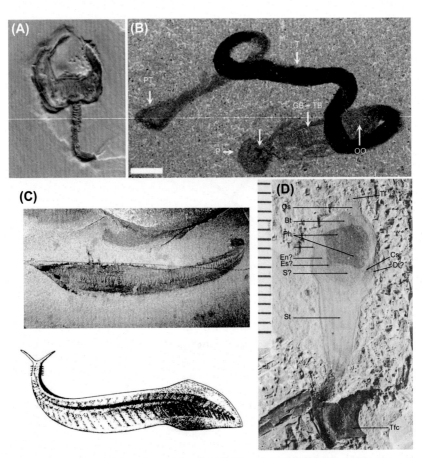

FIGURE 3.4   **Cambrian deuterostome fossils.** (A) The stem group echinoderm, *Cothurnocystis* *(From Erwin, D.H., Laflamme, M., Tweedt, S.M., Sperling, E.A., Pisani, D., Peterson, K.J., 2011. The Cambrian conundrum: early divergence and later ecological success in the early history of animals. Science 334, 1091–1097.)*. (B) Hemichordate acorn-worm fossil, *Spartobranchus tenuis (From Caron, J.B., Morris, S.C., Cameron, C.B., 2013. Tubicolous enteropneusts from the Cambrian Period. Nature 495, 503–506.)*. GB, gill bars; OO, esophageal organ; P, proboscis; PT, posterior trunk; T, trunk; TB, tongue bars. Scale bars represent 2 mm. (C) *Pikaia*, a fossil (upper) and a drawing (lower) *(From various sources.)*. (D) Tunicate ascidian fossil, *Cheungkongella (From Shu, D.G., Chen, L., Han, J., Zhang, X.L., 2001. An early Cambrian tunicate from China. Nature 411, 472–473.)*. Bt, buccal tentacles; Cs, cloacal siphon; Dt?, presumed degenerating tail; Os, oral siphon; En?, presumed endostyle; Es?, possible esophagus; Ph, pharynx; S?, presumed stomach; St, stem; T, tunic; Tf, tentacle-like fringe; Tfc, trilobite free cheek. *Question marks* indicate uncertainty. Scale bar, 1 mm.

phylum recognizes five extant classes, only a small fraction of past echinoderm diversity, which included approximately 20 classes in total. Early echinoderms were initially bilateral (see below), and all possessed a water-vascular system with tube feet. A pronounced diversification occurred at the class level during the Ordovician (Paul and Smith, 1984; Fig. 3.5).

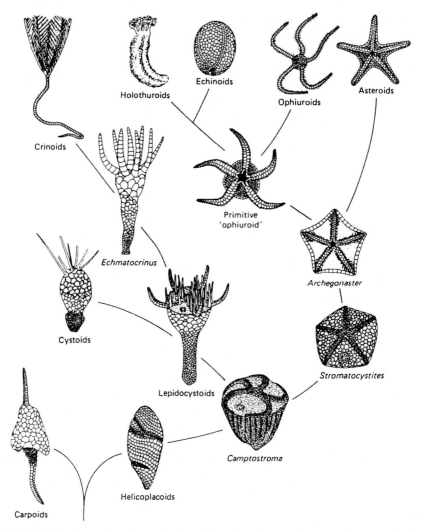

FIGURE 3.5 A possible scheme for the relationships of early echinoderms based on fossil records. *(From Paul and Smith, 1984. Biological Reviews of the Cambridge Philosophical Society. 59, 443–481.)*

Given the highly diversified body plans of adult echinoderms, extant animals have so far provided little insight into early origins of deuterostomes; however, their rich fossil record sheds some light on early deuterostome diversification. Some echinoderm fossils have obvious radial symmetry, but many others are strikingly asymmetric and some even show bilateral symmetry, suggestive of early stages in the development of this unusual body plan from a bilateral deuterostome ancestor. One particularly interesting group of asymmetrical fossils comprising mitrates and cornutes also likely possessed gill slits that appear homologous to hemichordate and chordate gill slits, further supporting the pandeuterostome homology of gill slits. This was discussed in Section 2.3.

3.3.2 Hemichordates

Hemichordates consist of two major groups: vermiform enteropneusts and minute, colonial, tube-dwelling pterobranchs (Section 1.3). Hemichordate phylogeny has long been problematic, not least because of the conjectural nature of any transitional form that might link the anatomically disparate enteropneusts and pterobranchs. Interrelationships among enteropneusts have also remained controversial. In addition, the hemichordate fossil record is almost entirely restricted to peridermal skeletons of pterobranchs, notably graptolites (Mitchell et al., 2013). Because of their low preservation potential, fossil enteropneusts are exceedingly rare and throw almost no light on either hemichordate phylogeny or the proposed harrimaniid–pterobranch transition.

However, a description of soft-bodied fossils of an enteropneust, *Spartobranchus tenuis* (Walcott, 1911), from the Middle Cambrian Burgess Shale [~520 million years (MY)] provides some insight into the relationship between enteropneusts and pterobranchs (Fig. 3.4B; Caron et al., 2013). The *S. tenuis* body is flexible, consisting of a short proboscis; a collar; and a narrow, elongate trunk terminating in a bulbous structure (Fig. 3.4B). The pharyngeal region accounts for approximately 10–20% of the total length. It is wider toward the mouth and tapers posteriorly (Fig. 3.4B). The dorsal branchial pharynx comprises thick and thin bars, interpreted as gill and tongue bars, respectively. Approximately one-quarter of specimens are associated with tubes that have a fibrous composition (see below).

S. tenuis is remarkably similar to extant harrimaniids but differs from all known enteropneusts in that it is associated with a fibrous tube that is sometimes branched. It is suggested that this is the precursor of pterobranch periderm and supports the hypothesis that pterobranchs are miniaturized and derived from an enteropneust-like worm. The presence of fossils of enteropneusts and pterobranchs in Middle Cambrian strata suggests that hemichordates originated at the onset of the Cambrian explosion. In addition, the finding that modern Atlantic acorn worms from the deep sea form tubes alternatively suggests that these fossils are similar to modern-day torquaratorids and that some behaviors have been conserved for more than 500 MY (Halanych et al., 2013).

3.3.3 Cephalochordates

The Burgess Shale and Chengjiang sites (Hou et al., 2007; Royal Ontario Museum, 2011) comprise several fossil metazoans that have been referred to as chordates because they show at least some indication of either a segmented body structure, a notochord, or gill slits. Segmentation of body musculature and a gill apparatus are often regarded as the "signature" of the chordates. Notably, this was the case for *Pikaia* (Fig. 3.4C), from the Burgess Shale, the body of which shows indications of a series of myomeres and a notochord. However, the head of *Pikaia* bears peculiar appendages (regarded as respiratory organs) and tentacles that have no apparent homology to vertebrate anatomy. Despite the exquisite preservation of numerous specimens of *Pikaia*, this long iconic "vertebrate ancestor" remains enigmatic. Opinions are still divided as to whether its affinities are to chordates or whether they represent convergent morphology with some protostomes (Conway–Morris and Caron, 2012).

In addition to *Pikaia*, Yunnanozoans (*Yunnanozoon* and *Haikouella*) from Chengjiang have also been classified as chordates (Chen et al., 2002a; Shu and Conway–Morris, 2003) because of their presumed notochord, segmented body musculature covered with a cuticle, and their vertebrate-like six pairs of gills. They have been variously considered stem hemichordates, stem cephalochordates, or stem vertebrates. The controversy over the stem-vertebrate and stem-deuterostome hypotheses (Mallat et al., 2003; Shu et al., 2003) reflects the difficulty in assessing the nature of the actual tissues and anatomical characters observed in fossils. Vetulicolans (*Vetulicola*, *Xidazoon*, *Didazoon*, and *Pomatrum*) from Chengjiang and the somewhat similar *Banffia* from the Burgess Shale display a bipartite structure, with a balloon-shaped, cuticle-covered head, laterally pierced by five presumed gill openings, and a flattened segmented tail. Again, the vetulicolan gill openings might suggest a stem deuterostome. On the other hand, the purported presence of an endostyle (a gland unique to chordates) suggests stem chordate affinity. *Cathaymyrus* from Chengjiang was described as "*Pikaia*-like." It has a worm-shaped body with a long series of myomeres and a distinct row of closely set pharyngeal slits that resemble those of cephalochordates.

However, an interesting speculation is that these fossils demonstrate some features interpreted as intermediate between those of cephalochordates and early vertebrates. New technology including high-resolution X-ray microtomography may disclose additional characters of these fossils, which may link them to extant chordates or enteropneusts.

3.3.4 Urochordates (Tunicates)

Because of the soft tissues of urochordates, it was once thought that they probably did not leave a fossil record. However, well-preserved fossils from Chengjiang indicate that *Cheungkongella* (Shu et al., 2001) (Fig. 3.4D) and *Shankou clava* (Chen et al., 2003) are tunicates. Both show the pharyngeal sac in the midst of the body (Fig. 3.4D). Interestingly, the thick tunic that covers the whole body is evident in both species. This suggests that tunicates had already emerged in the early Cambrian explosion and that they had acquired the tunic before the time of these fossils.

However, as a whole, all of these presumed chordates from the Cambrian, preserved mostly as soft-tissue imprints, provide scant information about their possible phylogenetic relationships. It is rather hard to infer the diversification of tunicate anatomy and morphology from fossils, much less their relationships to other cephalochordates and to vertebrates.

3.4 FOSSIL RECORDS OF VERTEBRATES

Advances in recent vertebrate paleontology have been discussed in the detailed reviews of Janvier (2015) and Brazeau and Friedman (2015). In addition, because the origin and evolution of vertebrates are not the main focus of this book, this topic will be only briefly described. Vertebrates exhibit, along with arthropods and echinoderms, many complex structures capable of producing fossils, which can be analyzed to reconstruct their relationships (eg, Ahlberg, 2001; Benton, 2005). Most of the anatomically useful vertebrate fossils appeared relatively late in the fossil record, approximately 470 MYA, suggesting that their emergence as fossils took more time compared with echinoderms and cephalochordates, presumably because of the evolution of their more complex body form. However, during the last decade, the fossil record of jawless vertebrates from 535 to 250 MYA has been enriched by the discovery of spectacular, well-preserved, soft-bodied fossils (sometimes known as Konservat-Lagerstätte). These fossils have added a new dimension, thanks to imaging techniques that allow the actual nature of preserved tissues to be determined as well as providing a better understanding of the processes involved in decay and fossilization, thereby avoiding overinterpretation (Sansom et al., 2010). It is now the consensus that early vertebrates emerged at the same time as invertebrates (Figs. 3.1 and 3.2). Distributions and patterns of interrelationships of the major Paleozoic jawless vertebrate groups and their extant relatives are shown in Fig. 3.6B.

FIGURE 3.6 **Fossils of early-stage vertebrates.** (A) The lower Cambrian agnathan vertebrate, *Haikouichthys ercaicunensis. (From Shu, D.G., Luo, H.L., Morris, S.C., Zhang, X.L., Hu, S.X., Chen, L., Han, J., Zhu, M., Li, Y., Chen, L.Z., 1999. Lower Cambrian vertebrates from South China. Nature 402, 42–46.)* (a) Entire specimen, anterior end to the right, posterior tip incomplete because of postmortem folding; scale bar = 5 mm. (b) Camera-lucida drawing of specimen, with certain features (notably structures interpreted as extrabranchial atria) to show interpretation. (c) Drawing of the entire body. (B) Distribution through geological time (*bold bars*), and patterns of interrelationships (*thin bars*) of the major Paleozoic jawless vertebrate groups and their extant relatives. *(From Janvier, P., 2015. Facts and fancies about early fossil chordates and vertebrates. Nature 520, 483–489.)* (C) The body plan of a hypothetical early vertebrate, emphasizing the pharynx, the dorsal nervous system, and myotomes. (From Romer, A.S., 1972. The vertebrate as a dual animal-somatic and visceral. Evolutionary Biology 6, 121–156.) (D) A hypothetical early vertebrate, based upon fossils. This reconstruction is a curious mix of rather vertebrate-like or even an ostracoderm-like head, and some cephalochordate characters, suggesting that the overall morphology of the common ancestor to cephalochordates and vertebrates was rather vertebrate-like. *(From Janvier, P., 2015. Facts and fancies about early fossil chordates and vertebrates. Nature 520, 483–489.)*

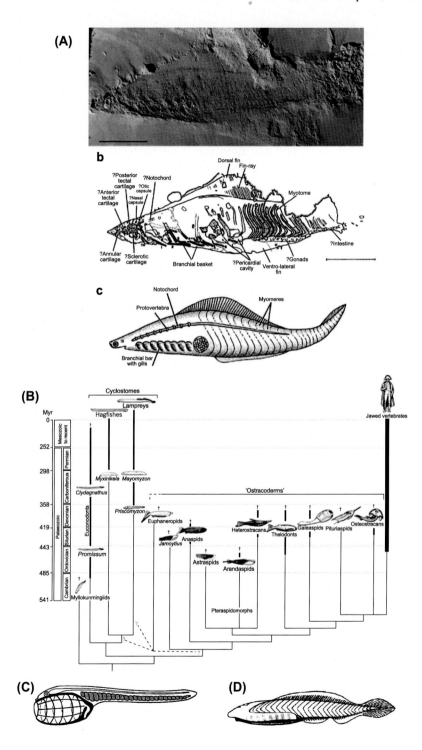

3.4.1 Cyclostome-Like Fossils

The concept of living cyclostome paraphyly, that lampreys are more closely related to gnathostomes than to hagfishes, was based only on phenotypic data derived from extant species. Certain ostracoderms (osteostracans and anaspids) were long considered to be most closely related to lampreys. More accurate character analyses showed that "ostracoderms" were, in fact, stem gnathostomes. The recent revival of cyclostome monophyly has no major bearing on their interrelationships. The fossil lamprey *Mayomyzon*, which was discovered in Carboniferous 300-MY-old rocks of the Mazon Creek Lagerstätte in Illinois, shows striking overall resemblance to a small modern lamprey. The image shows the outline of the body, the gill pouches, and the characteristic cartilages of the "tongue" apparatus. On the other hand, *Myxineidus* was referred to as a hagfish based only on the impression of two V-shaped rows of keratinous teeth that resemble those of living hagfishes. The Mazon Creek Lagerstätte has also yielded peculiar, presumed, soft-bodied jawless vertebrates, *Pipiscius* and *Gilpichthys*. The former has a lamprey-like oral funnel, and the latter shows possible impressions of sharp, nonmineralized teeth that resemble those of hagfishes. However, this interpretation remains controversial. Neither the fossil lampreys (eg, Bardack and Zangerl, 1968) nor the possible fossil hagfishes (eg, Bardack, 1991) show any clear indication of a mineralized skeleton.

3.4.2 Early Vertebrate Fossils

The myllokunmingiids, *Myllokunmingia* and *Haikouichthys* (Fig. 3.6A; Shu et al., 1999, 2003), from Chengjiang and the similar *Metaspriggina* (Conway–Morris and Caron, 2014) from the Burgess Shale look clearly fish-like, although the latter provides better information about the arrangement of gill bars and eye structure. Analyses of their characters suggest that myllokunmingiids are paraphyletic with *Myllokunmingia*, and that *Metaspriggina* is probably a stem vertebrate whereas *Haikouichthys* is a stem lamprey (Shu et al., 1999). By combining myllokunmingiid and *Metaspriggina* data, a better reconstruction of the most likely Cambrian vertebrates is possible — a jawless "fish" with a pair of large, anterodorsally facing camera eyes, a small median olfactory organ, five to seven pairs of gill arches, a stomach, a series of chevron-shaped myomeres, and a median fin web. These allow us to deduce hypothetical reconstructions of ancestral vertebrates by combining remotely resembling fossils and assembling imaged characters (Fig. 3.6 C and D). An atlas of fossil records of early bony vertebrates can be seen in the recent review of Brazeau and Friedman (2015).

3.4.3 Conodont Elements

One episode from the soft-bodied fossil record of vertebrates from the Cambrian to the Late Triassic (~530–200 MY) is the saga of the conodont, which has a small head with large, paired eyes; a mouth or pharynx containing many denticles; an elongated, eel-shaped body with chevron-shaped myomeres; and a small caudal fin supported by possibly cartilaginous rods. Conodont elements were first discovered in 1856, and they are minute fossils resembling spines, combs, or vertebrate teeth. Because some of their body parts are formed of calcium phosphate, similar to vertebrate teeth, they have been considered the earliest examples of mineralized vertebrate skeletons. This inspired the "inside-out" hypothesis—the possibility that teeth evolved independently of the vertebrate dermal skeleton and before the origin of jaws (Gee, 1996). A recent analysis using synchrotron radiation X-ray tomographic microscopy failed to support the putative homology of euconodont crown tissue and vertebrate enamel (Murdock et al., 2013). These tissues evolved independently and convergently. Now it seems likely that the last common ancestor of conodonts and jawed vertebrates probably lacked mineralized tissues. Teeth seem to have evolved through the extension of odontogenic competence from the dermis to internal epithelia soon after the origin of jaws. Current data suggest that euconodonts might be stem vertebrates; stem cyclostomes; or, less likely, stem lampreys or stem hagfishes.

3.5 CONCLUSIONS

Recent metazoan paleontology has dramatically advanced our understanding of the origin of metazoan body plans with the discovery of fossils, not only of hard tissues but also of soft body parts, as well as introduction of sophisticated new imaging and analytical techniques. The metazoan fossil record tells us at least two astonishing things. First, metazoans did not emerge gradually from those of lower morphological complexity (eg, sponges) to those with higher complexity (eg, chordates). Instead, almost all taxa appear to emerge suddenly in a very short time of period. Second this occurred during the Cambrian Period (from 541 to 485 MYA) and is therefore called the Cambrian explosion. We now have well-conserved fossils of all five deuterostome taxa, all of which emerged approximately 525 MYA. These fossils look very similar to extant deuterostomes. We must consider chordate origins and evolution with these facts in mind. In other words, each of these taxa already has a 500-MY history; therefore comparisons of body plans within a given taxon may be reasonable, but comparisons between taxa require caution. As discussed in later chapters, one significant issue in understanding the evolution of body plans is the relationship between cephalochordates and vertebrates. There are several intriguing fossils that resemble lancelets and early vertebrates. However, the path of vertebrate evolution is gradually becoming clearer. Further investigation of deuterostome fossils using more sophisticated techniques may shed light on the evolution of chordates.

Chapter 4

Molecular Phylogeny

Traditional metazoan phylogeny was mainly based on modes of embryogenesis, similarities of larval and adult morphology and anatomy, physiological characters, and fossil records. In contrast, molecular phylogeny is based on discrete, presumably neutral changes in molecules; thus it is more objective and powerful. What does modern molecular phylogeny tell us about the phylogenetic relationships of deuterostomes?

Traditional phylogeny and evolution of metazoans have been studied and discussed primarily on the basis of modes of embryogenesis, similarities of larval and adult morphology and anatomy, physiological traits, and fossil records (Willmer, 1990; Gee, 1996; Hall, 1999; Nielsen, 2012; Chapters 1–3). This approach has established a robust scheme of metazoan phylogeny, including diploblasts, triploblasts (bilaterians), protostomes, deuterostomes, and chordates. However, discussions have sometimes become controversial because of the subjectivity of researchers in selecting or emphasizing certain attributes to determine relationships (eg, see Section 1.2.3 for the history of recognition of chordates). On the other hand, inferring metazoan phylogeny on the basis of molecular characters is more objective. Although many molecular changes are deleterious, and some are beneficial, most molecular changes, such as substitutions of nucleotides in genomic DNA and amino acids of proteins, are believed to be neutral so that the progression of sequence changes reflects the length of time since the divergence of two or more taxa. Molecular phylogeny of deuterostomes began by comparisons of 18S ribosomal RNA/DNA (rRNA/rDNA) sequences (Wada and Satoh, 1994; Turbeville et al., 1994; Halanych, 1995) and later on protein-coding gene sequences (Dunn et al., 2008; Philippe et al., 2009). However, problems can arise depending upon which molecules researchers compare and what bioinformatics tools they use; therefore caution must still be exercised. For example, a phylogeny based on 18S rDNA does not always coincide with one based on mitochondrial data (eg, Perseke et al., 2013).

Technical innovations, especially polymerase chain reaction (PCR), have also greatly advanced molecular phylogeny. Using highly conserved sequence regions as primers, long reads of 18 and 28S rDNAs have been obtained and compared. RNA-sequencing technology with new-generation sequencers also provides transcriptomic data for molecular phylogeny. This and other technical innovations make molecular phylogeny easier, allowing even species of minor taxa to be studied.

Chordate Origins and Evolution. http://dx.doi.org/10.1016/B978-0-12-802996-1.00004-3

In addition, new bioinformatics tools with improved algorithms for sequence comparisons have contributed to make inferences more accurate. Today, nuclear genome sequences and protein amino acid sequences are frequently used for comparisons. In addition to quantitative differences in the sequences, qualitative differences, such as synteny (Hox cluster or code for example; see Section 6.3), indels, intron-exon constitution of genes (see Section 5.4), addition and/or loss of microRNAs (miRNAs; see Section 4.5), and repetitive elements have been applied to phylogeny inference.

4.1 MOLECULAR PHYLOGENY OF METAZOANS

The introduction and/or application of molecular phylogenetic methods to metazoans commenced in the 1980s. This was delayed compared with other organisms, such as prokaryotes, fungi, and plants, because every metazoan taxon has distinctive features, including adult anatomy and mode of embryogenesis. One of the most striking results of metazoan molecular phylogeny is that protostomes are divided into two major groups: the Lophotrochozoa (those having lophophore or trochophore larvae and sometimes called Spiralia, in a wider sense) and the Ecdysozoa (those that exhibit molting; Fig. 1.1; eg, Aguinaldo et al., 1997; Halanych et al., 1995). The former includes annelids, molluscs, brachiopods, and flat worms, and the latter includes nematodes and arthropods. This phylogenic interpretation is different from the traditional one in which protostomes were subdivided mainly on the basis of the mode of formation of the body cavity (or coelom). Acoelomates such as the platyhelminthes have no distinct body cavity; pseudocoelomates, such as nematodes, have poorly developed body cavities; and coelomates, such as annelids, molluscs, and arthropods, have distinct body cavities.

More recent molecular phylogeny, including microbenthic lineages, suggests that the "Lophotrochozoa," including the Mollusca, Annelida, and Brachiopoda, is one large group of "Spiralians," also including the Platyhelminthes, Gastrotriches, and Rotifera (Laumer et al., 2015). Although it is not my intent to become mired in terminology here, because more than half of all protostomes are grouped as Spiralians, the mode of embryogenesis is critical to infer their ancestry. In addition, ecdysozoans are unique in terms of metazoan diversification. Ecdysozoan gene family composition is different from those of Spiralians and invertebrate deuterostomes (see Fig. 5.3).

However, the magnitude of molecular changes is not always comparable; in some lineages molecules change rather quickly whereas in others at the same taxonomic level they change very slowly. An extraordinarily high rate of nucleotide substitution appears as long branches in a phylogenetic tree, causing a distortion of the tree called a "long-branch attraction problem." Because of long-branch attraction problems, several aspects of metazoan molecular phylogeny continue to be controversial. Positioning of ctenophores at the base of metazoans (Dunn et al., 2008; Hejnol et al., 2009; Ryan et al., 2013; Moroz et al., 2014), the affinity of Xenacoelomorpha to deuterostomes (see Section 4.4), and the enigmatic position of appendicularians among tunicates are cases in point.

In addition, there are several groups for which phylogenic positions are still enigmatic, such as the mesozoans and chaetognaths.

4.2 MOLECULAR PHYLOGENY OF DEUTEROSTOME TAXA

A recent molecular phylogeny paid special attention to deuterostome relationships (Fig. 4.1; Simakov et al., 2015). The tree was constructed by comparing approximately 500,000 amino acid positions in 1565 protein families with single-copy orthologs in 53 metazoan species with 30 sequenced genomes, in which presence-absence characters for introns and coding indels were also incorporated. This and previous molecular phylogenies have unambiguously demonstrated at least two major phylogenetic issues. First, the division of deuterostomes into two major groups, Ambulacraria and Chordata, and second, cephalochordates diverged first among the chordate lineages.

4.2.1 Ambulacraria and Chordata: Two Major Groups of Deuterostomes

Although an early phase of deuterostome molecular phylogeny using 18S ribosomal RNA (rRNA) failed to provide clear resolution of relationships among the five taxa because of a long-branch attraction problem caused by a high substitution rate among tunicate sequences (Wada and Satoh, 1994; Turbeville et al., 1994), the tree grouped echinoderms and hemichordates. Halanych et al. (1995) confirmed this result and suggested the name "Ambulacraria," originally proposed by Metchnikoff (1881; see Section 1.2.2). This grouping has been supported by various molecular phylogenetic analyses. With similarities in embryogenesis (dipleurula larvae and coelomic systems; Chapter 6), it is now the general consensus that echinoderms and hemichordates constitute a natural grouping, the Ambulacraria, and that they share common ancestry (Cameron et al., 2000; Swalla and Smith, 2008; Brown et al., 2008; Satoh, 2008).

On the other hand, molecular phylogeny using improved methods to carefully incorporate tunicate sequences supports the monophyly of chordates (see below). The presence of a notochord and other characters also unites urochordates, cephalochordates, and vertebrates as chordates. Since then, the monophyly of deuterostomes, ambulacrarians, and chordates has been supported by various sets of molecular data; thus it appears that from a common ancestor deuterostomes diverged first into two major groups, the Ambulacraria and the Chordata.

4.2.2 Cephalochordates Came First Among Chordates and Then Urochordates + Vertebrates as Olfactores

As discussed in Section 2.4, phylogenetic relationships among three chordate taxa have been vigorously discussed from the point of view of vertebrate origins, in which the earlier divergence of the urochordates was the prevailing view. A sophisticated

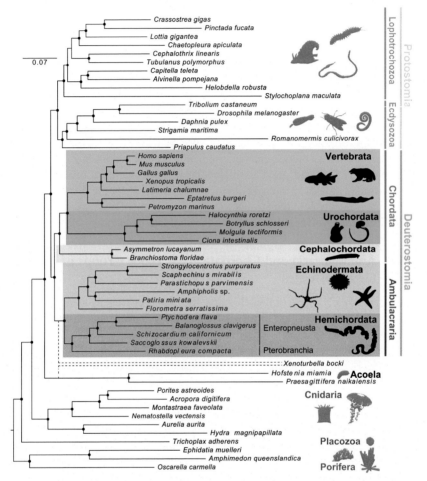

FIGURE 4.1 Molecular phylogeny of deuterostome taxa within the metazoan tree. Echinoderms are shown in *light green*, hemichordates in *magenta*, cephalochordates in *yellow*, urochordates in *dark green*, and vertebrates in *blue*. This maximum-likelihood tree was obtained with a supermatrix of 506,428 amino acid residues gathered from 1564 orthologous genes, using a Γ+LG model partitioned for each gene. *Plain circles* at nodes denote maximal bootstrap support. (*From Simakov, O., Kawashima, T., Marlétaz, F., Jenkins, J., Koyanagi, R., Mitros, T., Hisata, K., Bredeson, J., Shoguchi, E., Gyoja, F., Yue, J.X., Chen, Y.C., Freeman, R.M., Sasaki, A., Hikosaka-Katayama, T., Sato, A., Fujie, M., Baughman, K.W., Levine, J., Gonzalez, P., Cameron, C., Fritzenwanker, J.H., Pani, A.M., Goto, H., Kanda, M., Arakaki, N., Yamasaki, S., Qu, J., Cree, A., Ding, Y., Dinh, H.H., Dugan, S., Holder, M., Jhangiani, S.N., Kovar, C.L., Lee, S.L., Lewis, L.R., Morton, D., Nazareth, L.V., Okwuonu, G., Santibanez, J., Chen, R., Richards, S., Muzny, D.M., Gillis, A., Peshkin, L., Wu, M., Humphreys, T., Su, Y.H., Putnam, N.H., Schmutz, J., Fujiyama, A., Yu, J.K., Tagawa, K., Worley, K.C., Gibbs, R.A., Kirschner, M.W., Lowe, C.J., Satoh, N., Rokhsar, D.S., Gerhart, J., 2015. Hemichordate genomes and deuterostome origins. Nature 527, 459–464.)*

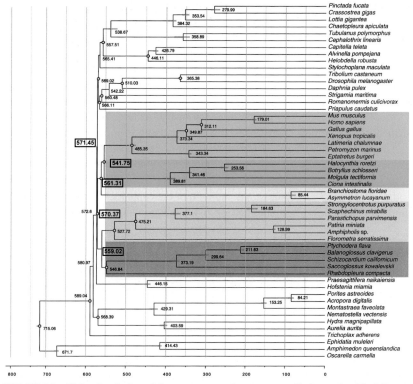

FIGURE 4.2 **Molecular dating of deuterostome and metazoan radiations using PhyloBayes.** Echinoderms are shown in *light green*, hemichordates in *magenta*, cephalochordates in *yellow*, urochordates in *dark green*, and vertebrates in *blue*. *Yellow circles* at some nodes indicate calibration dates from the fossil record. Bars are 95% confidence intervals derived from posterior distributions. This analysis suggests that deuterostomes emerged at ~570 MYA. (*From Simakov, O., Kawashima, T., Marlétaz, F., Jenkins, J., Koyanagi, R., Mitros, T., Hisata, K., Bredeson, J., Shoguchi, E., Gyoja, F., Yue, J.X., Chen, Y.C., Freeman, R.M., Sasaki, A., Hikosaka-Katayama, T., Sato, A., Fujie, M., Baughman, K.W., Levine, J., Gonzalez, P., Cameron, C., Fritzenwanker, J.H., Pani, A.M., Goto, H., Kanda, M., Arakaki, N., Yamasaki, S., Qu, J., Cree, A., Ding, Y., Dinh, H.H., Dugan, S., Holder, M., Jhangiani, S.N., Kovar, C.L., Lee, S.L., Lewis, L.R., Morton, D., Nazareth, L.V., Okwuonu, G., Santibanez, J., Chen, R., Richards, S., Muzny, D.M., Gillis, A., Peshkin, L., Wu, M., Humphreys, T., Su, Y.H., Putnam, N.H., Schmutz, J., Fujiyama, A., Yu, J.K., Tagawa, K., Worley, K.C., Gibbs, R.A., Kirschner, M.W., Lowe, C.J., Satoh, N., Rokhsar, D.S., Gerhart, J., 2015. Hemichordate genomes and deuterostome origins. Nature 527, 459–464.*)*

analysis incorporating orthologous amino acid sequences of appendicularians and cephalochordates clearly demonstrated that within the chordate clade, cephalochordates diverged first, leaving urochordates and vertebrates to form a sister group (Delsuc et al., 2006), sometimes called the "Olfactores" (Section 1.2.4). This relationship has been supported by further analyses that include different taxa and larger quantities of better molecular data (Bourlat et al., 2006; Putnam et al., 2008) (Figs. 4.1 and 4.2). That is, debates on evolutionary scenarios in relation to sedentary and free-living ancestors have now been resolved. The consensus now is that chordate ancestors were free-living, similar to lancelets (Fig. 2.5C).

The notion of the unity of urochordates and vertebrates and with cephalochordate having diverged first among chordates receives strong qualitative support from studies of cadherin structure (Fig. 4.3; Oda et al., 2002). The classic cadherin family of vertebrate proteins is characterized by Ca^{2+}-dependent cell–cell adhesion molecules with Ca^{2+}-binding repeats (ECs) in the extracellular region and a cytoplasmic domain (CP) with an evolutionarily conserved β-catenin–binding sequence. Urochordate cadherins have a domain composition identical to those of vertebrate cadherins (Fig. 4.3). On the other hand, classic cadherins of nonchordate deuterostomes contain ECs, a domain unique to nonchordate classic cadherins, a cysteine-rich epidermal growth factor-like domain (CE), a laminin G domain (LG), and a CP (Fig. 4.3). In contrast, the cadherin of the cephalochordate amphioxus is unique to chordates in containing LG, CE, and CP domains (Fig. 4.3).

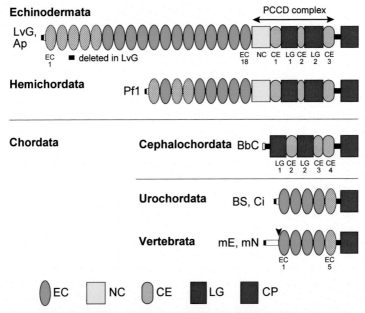

FIGURE 4.3   Modification of cadherins during deuterostome evolution. Comparison of the primary structures of sea urchin LvG-cadherin, hemichordate Pf1, cephalochordate BbC, urochordate BS and Ci, and vertebrate (mouse) E-cadherin. Signal sequences and transmembrane segments are indicated by *solid black boxes*. The arrowhead shown for mouse E-cadherin indicates a proteolytic cleavage site that is utilized in maturation of this protein. Domain abbreviations: *EC*, cadherin extracellular repeat; *NC*, nonchordate classic cadherin-specific domain; *CE*, cysteine-rich epidermal growth factor-like domain; *LG*, laminin G-like domain; *CP*, cytoplasmic domain; *PCCD complex*, primitive classic cadherin domain complex. (*Modified from Oda, H., Wada, H., Tagawa, K., Akiyama-Oda, Y., Satoh, N., Humphreys, T., Zhang, S.C., Tsukita, S., 2002. A novel amphioxus cadherin that localizes to epithelial adherens junctions has an unusual domain organization with implications for chordate phylogeny. Evolution and Development 4, 426–434.*)

More support for tunicate and vertebrate unity comes from opsin molecules. Invertebrates and vertebrates use photoreceptor cells with different morphologies and electrophysiological responses (eg, Willmer, 1990). Eyes of insects and molluscs use rhabdomeric photoreceptor cells, which depolarize in response to light. In contrast, vertebrate eyes use ciliary photoreceptor cells, which hyperpolarize in response to light. Photoreceptor cells of ascidian larvae are ciliary and hyperpolarize in response to light, indicating that urochordate photoreceptor cells are of the vertebrate type (Eakin and Kuda, 1971; Gorman et al., 1971; Kusakabe and Tsuda, 2007). Opsins are G-protein–coupled receptor proteins of 30–50 kDa. More than 1000 members of the opsin family have been categorized into 7 subfamilies corresponding to functional classifications within the family (Terakita, 2005). *Ci-opsin1*, an opsin gene from *Ciona intestinalis*, is expressed in the larval ocellus (Kusakabe et al., 2001). *Ci-opsin* is homologous to vertebrate opsins. In contrast, cephalochordates contain rhabdomeric photoreceptor-like opsins, but not true vertebrate-type opsins. This suggests that after diversification of the cephalochordate and urochordate/vertebrate lineages, the clade comprising urochordates and vertebrates developed opsins specific to this lineage (Kusakabe and Tsuda, 2007).

The original unity of urochordates and vertebrates as Olfactores was based upon similarities in extensive pharyngeal remodification leading to the formation of new structures that are not found in cephalochordates (Jefferies et al., 1996). Similarity in the extent of developmental novelties has been revealed by several studies using *Ciona* embryos and larvae (Chapter 10). This novel view of chordate phylogeny has become the consensus because a great variety of data from different disciplines supports arguments for them.

4.2.3 Timing of the Emergence of Deuterostome Groups

The estimated emergence times of deuterostomes extrapolated based on paleontology and molecular phylogeny differ considerably depending upon what kind of data researcher used for the analyses (eg, Wray et al., 1996; Blair and Hedges, 2005; Peterson et al., 2005, 2008; Erwin et al., 2011; Simakov et al., 2015). Fig. 4.2 shows a new metazoan phylogeny using a relaxed molecular clock to infer the timing of metazoan emergence and is based on an analysis similar to that in Fig. 3.2, but it incorporates more molecular data (Simakov et al., 2015). This tree estimates the divergence of deuterostomes and protostomes at approximately 571 million years ago (MYA). Ambulacrarians and chordates presumably diverged approximately 1 million years later (570 MYA). Echinoderms and hemichordates separated approximately 559 MYA, whereas cephalochordates diverged among the chordate taxa at approximately 561 MYA. On the basis of more paleontological data, Fig. 3.2 shows earlier emergence times than Fig. 4.2 (Erwin et al., 2011). According to Erwin et al. (2011), the divergence time of deuterostomes and protostomes was approximately 670 MYA. Ambulacrarians and chordates parted company approximately 660 MYA. Echinoderms and hemichordates diverged approximately 600 MYA, and the three chordate groups separated approximately 650 MYA.

Although the two analyses differ by approximately 100 million years, they suggest similar affinities among taxa. The tree clearly suggests that (1) all five deuterostome taxa emerged in a relatively short time before the Cambrian Period and (2) the divergence of chordates was slightly earlier than that of the ambulacrarians. This gives us some insight into the evolution of deuterostomes. That is, if we trust the tree, chordate evolution was independent of ambulacrarian evolution. Although chordates share a common ancestor with ambulacrarians, they did not originate downstream of ambulacrarians. The Cambrian origin of hemichordates, as suggested by recent fossils, is evident in this tree (Fig. 4.2).

4.3 RELATIONSHIPS WITHIN EACH DEUTEROSTOME PHYLUM

Molecular phylogenic studies have also clarified or supported relationships of members within higher taxa.

4.3.1 Echinoderms

As discussed in Section 1.3.1, there are five distinct classes of echinoderms, including the Crinoidea (sea lilies and feather stars), the Asteroidea (sea stars), the Ophiuroidea (serpent or brittle stars), the Echinoidea (sea urchins and sand dollars), and the Holothuroidea (sea cucumbers; Fig. 1.2A). Well-preserved echinoderm fossils suggest that crinoids diverged first from a common ancestor, and that they are the most distant from the four other echinoderm classes, in which asteroids, ophiuroids, echinoids, and holothuroids emerged in this order (Fig. 3.5). Molecular phylogenies including comparison of 18S rRNA sequences support this notion (eg, Wada and Satoh, 1994). This provides a good example of paleontology supported by molecular phylogeny.

4.3.2 Hemichordates

Hemichordate phylogeny has long remained problematic, not the least because the nature of any transitional form that might serve to link the anatomically disparate enteropneusts and pterobranchs is conjectural (Fig. 1.3). Hence, interrelationships have also remained controversial (Cannon et al., 2009, 2013; Osborn et al., 2012; Peterson et al., 2013; Zeng and Swalla, 2005). For example, pterobranchs have sometimes been compared to ancestral echinoderms and regarded as an independent phylum (Nielsen, 2012). Some molecular data identified enteropneusts as paraphyletic and harrimaniids as the sister group of pterobranchs (Cameron et al., 2000). Recent molecular phylogenies suggest that enteropneusts are probably basal among hemichordates, contrary to previous views, but otherwise they provide little guidance as to the nature of primitive hemichordates.

A molecular phylogeny using decoded genomes of two acorn worms and RNA data from a rhabdopleurid pterobranch clearly assigns pterobranch hemichordates as the sister group to enteropneusts rather than within them (Fig. 4.1). Phylogenetic analysis implies that genomic traits shared by chordates and ambulacrarians can be attributed to the last common deuterostome ancestor. Studies of pterobranch embryonic development are essential to understand hemichordate evolution. Another interesting thing is that the divergence time of *Saccoglossus kowalevskii* (Harrimaniidae; Atlantic, North America) and *Ptychodera flava* (Ptychoderidae; Pacific, pan-tropical) was estimated to be approximately 370 MYA (Fig. 4.2). Although *S. kowalevskii* develops within days directly into a juvenile worm having these traits, *P. flava* develops indirectly through a feeding larva that metamorphoses to a juvenile worm after months in a planktonic larval state (Fig. 1.3C). These two hemichordates provide clues to study hypotheses of larval evolution by comparing direct-developing and indirect-developing acorn worms, which achieve remarkably similar adult forms by distinctly different embryonic routes. This raises the question of whether the direct mode of embryogenesis is primitive or advanced.

4.3.3 Cephalochordates

The Cephalochordata is a very small phylum, comprising only approximately 35 species. It consists of two genera, *Branchiostoma*, with paired gonads, and *Epigonichthys* with right-side gonads (Nohara et al., 2005; Kon et al., 2007; Somorjai et al., 2008). It has been said that cephalochordate species have large population sizes. *Branchiostoma floridae* and *Asymmetron lucayanum* (both Family Branchiostomidae) diverged rather recently, only approximately 85 MYA, in contrast to the very early divergence time of their lineage (~561 MYA). This suggests that their common lifestyle is not subject to diversifying selection (Chapter 7).

4.3.4 Urochordates (Tunicates)

Traditionally tunicates are thought to comprise three classes: Ascidiacea, Thaliacea, and Appendicularians. The ascidian class includes Stolidobranchia, Phlebobranchia, and Aplousobranchia. Tunicates have diverged into approximately 3000 extant species, which exhibit a wide variety of adult morphologies and lifestyles. In addition, a long-branch attraction problem due to a high rate of molecular changes that might cause variable adult morphology makes phylogenetic relationships among tunicate classes enigmatic (Swalla et al., 2000; Turon and López-Legentil, 2004). This is typical of long-branch attraction problems in metazoan molecular phylogeny.

The traditional view of three tunicate classes is now challenged by molecular data. Molecular phylogenies using 18S rDNAs and mitochondrial gene sequences have shown that the Thaliacea is included in the Ascidiacea as a sister group to the Phlebobranchia (Stach and Turbeville, 2002). In addition, the Aplousobranchia, such as *Didemnum* and *Clavelina*, forms a strongly supported monophyletic group sister to the Stolidobranchia and Phlebobranchia. Furthermore, the Appendicularia (Larvacea) nests within the Ascidiacea as the sister taxon of the Aplousobranchia (Stach, 2008), although some insist upon the basal position of appendicularians among tunicates (Satoh, 2009). However, molecular phylogeny of ascidians provides evidence regarding at least two questions of ascidian biology. First, colonialism evolved independently in various lineages of ascidians (Wada et al., 1992); and second, direct development has also arisen several times independently among the families Molgulidae and Styelidae (Hadfield et al., 1995). Data regarding gene order in the mitochondrial genome sometimes provide useful information to infer relationships among tunicates (Yokobori et al., 2005).

4.3.5 Vertebrates

Vertebrates comprise agnathans and gnathostomes. They have extensive fossil records, thereby enabling us to deduce the evolutionary history of jawed vertebrates in particular. Debates have centered on the relationship of lampreys and hagfish to each other and to other vertebrate groups. This question is also related to the question when during the evolution of vertebrates two rounds of genome-wide gene duplication occurred—before or after the divergence of cyclostomes. Cyclostomes and other fishes diverged approximately 485 MYA, and it took another 140 MY for divergence of lampreys and hagfish (Fig. 4.2). Separation of agnathans and gnathostomes, and subsequent divisions among gnathostomes, probably occurred between 370 and 312 MYA.

One of the most interesting suggestions from molecular phylogeny is the placement of birds within reptiles (Hedges and Poling, 1999; Alföldi et al., 2011). In addition, the classification of gnathostomes remains one of vigorous debate, although vertebrate phylogeny is not the primary focus of this book.

4.4 XENACOELOMORPHA

I will not devote much attention to the Xenoturbellida and Acoelomorpha, mainly because these animals are not directly associated with the origins of chordates, although they are pertinent to a discussion of deuterostome origins. *Xenoturbella bocki* is a centimeter-long, marine "turbellariform" worm found in 1950 (Fig.4.4A(d); Telford, 2008). The outer surface is covered with a thick, ciliated epidermis. It also possesses a thick, extracellular matrix; a layer of strong muscle; and a sac-shaped gut with a mid-ventral mouth opening. The xenoturbellarian nervous system is a diffuse, intraepithelial nerve net without any special concentration. There are no other obvious internal structures. On the other hand, until recently, acoels were recognized as primitive platyhelminths, and nemertodermatids comprised a primitive, but independent phylum (Fig. 4.4A(e)). On the basis of morphological similarities, *Xenoturbella* has been recognized as closely related to acoels and is sometimes placed in the clade Xenacoelomorpha.

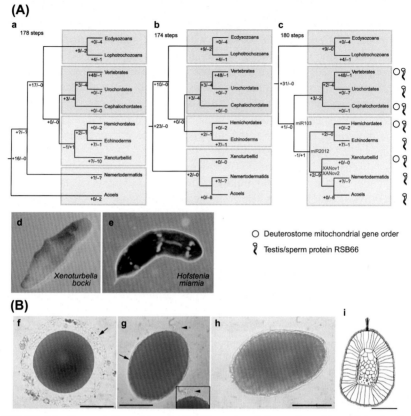

FIGURE 4.4 *Xenoturbella* **and acoels.** (A) Alternative phylogenetic positions of the Acoela, Nemertodermatida, and Xenoturbellida with implied evolution of different characters. (a) Tree based on previous studies showing positions of nemertodermatids, acoels, and *Xenoturbella*. (b) Tree based on other analyses. (c) Tree based on Philippe et al. (2011). Protein RSB66 and deuterostome mitochondrial gene order are also indicated. microRNAs representing possible synapomorphies of the Deuterostomia, Xenambulacraria, and Xenacoelomorpha are shown in *red*. The minimum number of total steps to explain microRNA distribution is shown above the trees. Losses and acquisitions of microRNAs are shown on each branch. (d) *Xenoturbella bocki* and (e) *Hofstenia miamia* are shown in the left corner. *(From Philippe, H., Brinkmann, H., Copley, R.R., Moroz, L.L., Nakano, H., Poustka, A.J., Wallberg, A., Peterson, K.J., Telford, M.J., 2011. Acoelomorph flatworms are deuterostomes related to Xenoturbella. Nature 470, 255–258.).* (B) Embryogenesis of *X. bocki* is similar to that of acoels. (f) Unfertilized egg with jelly coat *(arrows)*. (g) Fertilized egg with fertilization membrane *(arrow)*. Arrowheads and *insert* indicate sperm. (h) Gastrula and (i) a sketch thereof showing inside of the gastrula. *(From Nakano, H., Lundin, K., Bourlat, S.J., Telford, M.J., Funch, P., Nyengaard, J.R., Obst, M., Thorndyke, M.C., 2013. Xenoturbella bocki exhibits direct development with similarities to Acoelomorpha. Nature Communications 4, 1537.)*

It was molecular phylogeny that evoked an intense discussion about these animals. The first molecular phylogeny with rRNA suggested that *Xenoturbella* belongs to the Mollusca. Later, it was shown that this result is because *Xenoturbella* eats mostly bivalves, and this latter study placed *Xenoturbella* as sister group of the Ambulacraria (Fig. 4.4A(a); Bourlat et al., 2003, 2006). In addition, further molecular phylogenies using miRNA data and the gene *Rsb66* placed *Xenoturbella* together with acoels and nemertodermatids, a more basal position, as a sister group to all other deuterostomes (Fig. 4.4A(c); Philippe et al., 2011). However, because of evolutionary molecular changes that occurred in xenacoelomorphs, analyses with molecular data did not always achieve a consensus. The molecular phylogeny shown in Fig. 4.1 suggests that acoels are positioned as a bilaterian sister group, although a deuterostome placement cannot be ruled out because a very long-branch attraction damages tree topology (Fig. 4.4A(b)).

As discussed in Section 5.4, deuterostomes possess a well-organized Hox code, although several modifications are evident in certain groups, especially in urochordates. In contrast, the Hox code of the Xenacoelomorpha is very primitive (Cook et al., 2004). Although it has been studied solely by PCR amplification survey, *X. bocki* contains only an anterior *Hox1*, central *HoxM1*, *HoxM2*, and *HoxM3* genes, and a posterior *HoxP* gene (Fritzsch et al., 2008). On the other hand, acoels have only three *Hox* genes: one anterior, one central, and one posterior. Nemertodermatida have one posterior and two central *Hox* genes (Jiménez-Guri et al., 2006). In addition, as discussed in Section 6.2, the pharyngeal gene cluster illustrates another shared character of deuterostomes in relation to the pharynx, and the *Pax1/9* gene is a major component of the cluster. *Pax1/9* was not found in the Xenacoelomorpha.

Furthermore, the mode of embryogenesis is completely different between Xenacoelomorpha and deuterostomes (Fig. 4.4B). Embryogenesis of acoels is direct. Radial cleavage forms the blastula, and gastrulation occurs via delamination of blastomeres (Fig. 4.4B). The mode of embryogenesis of *X. bocki* is similar to that of acoels (Nakano et al., 2013) and different from that of ambulacrarians.

If xenacoelomorphs are basal to deuterostomes, then deuterostomes should be reconsidered from their base. Deuterostomes are defined on the basis of embryogenesis, in which the blastopore becomes the anus and the mouth develops secondarily from a new opening almost opposite the anus (Table 1.1). Adults share pharyngeal slits. If we include xenacoelomorphs within deuterostomes, then how do we define deuterostomes?

4.5 MICRORNAS

miRNAs encode approximately 22-nucleotide, noncoding regulatory RNAs that affect translation of target mRNAs, ultimately contributing to the maintenance of cellular homeostasis and cellular identity and to the robustness of developmental programs (eg, Peter and Davidson, 2015). Kevin Peterson and colleagues have shown that miRNAs seem to have been continuously added to eumetazoan genomes through time, with very little secondary loss in most taxa

(Peterson et al., 2009; Tarver et al., 2013; Erwin et al., 2011). The expansion in the number of miRNA families is correlated with an increase in the number of cell types and the morphological complexity of animals, especially in vertebrates, and that a decreased number of miRNA families is associated with morphological simplification (Fig. 4.5). According to their analyses, *Xenoturbella* and acoel flat worms are characterized by extensive secondary loss of miRNA complements compared with ambulacrarian deuterostomes. Although deuterostomes and xenacoelomorphs may share common ancestry, as mentioned earlier, this book does not discuss xenacoelomorphs in the context of deuterostome evolution leading to chordates.

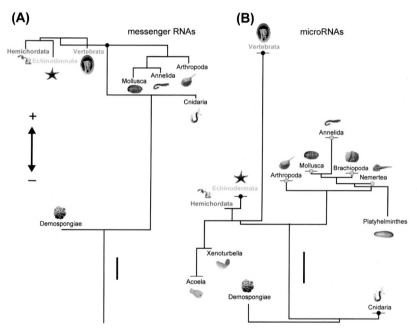

FIGURE 4.5   **Acquisition and secondary loss of mRNAs (A) and microRNAs (B) in selected taxa.** One hundred and thirty-one representative transcription factors and signaling ligands were coded for eight metazoan taxa and mapped onto a widely accepted metazoan topology. The length of the branch represents the total number of mRNA genes acquired minus those that were lost (scale bar represents 10 genes total). Much of the developmental mRNA toolkit was acquired before the last common ancestor of cnidarians and bilaterians. This is in contrast to the microRNA repertoire that displays extensive acquisition of microRNAs in the bilaterian stem lineage after it split from cnidarians. All 139 miRNA families known from 22 metazoan species were coded, and similar to the mRNA figure (A), the length of the branch represents the total number of microRNA genes acquired at that point minus those that were lost secondarily (scale bar represents 10 genes total). Increases in morphological complexity are correlated with increases in the microRNA toolkit, and secondary simplifications in morphology correlate with a relatively high level of secondary microRNA loss. *(From Erwin, D.H., Laflamme, M., Tweedt, S.M., Sperling, E.A., Pisani, D., Peterson, K.J., 2011. The Cambrian conundrum: early divergence and later ecological success in the early history of animals. Science 334, 1091–1097.)*

4.6 CONCLUSIONS

Molecular phylogeny is a powerful method to resolve phylogenetic questions regarding metazoans. The method is objective because it is based on comparisons of genomic DNA and protein amino acid sequences. Qualitative genetic and genomic traits have also offered insights into various aspects of deuterostome relationships. Recent molecular phylogenetic, genomic, and evolutionary developmental studies of deuterostomes have unambiguously demonstrated that echinoderms and hemichordates form a clade (Ambulacraria) and that cephalochordates, urochordates, and vertebrates form another (Chordata). In addition, within the chordate clade, cephalochordates diverged first, and urochordates and vertebrates form a sister group (sometimes called Olfactores). This novel consensus view of deuterostome taxonomy and phylogeny is robust because a great variety of data from different disciplines supports it. In addition, molecular phylogeny suggests that all five deuterostome taxa emerged at nearly the same time, around 570 MYA. Furthermore, an estimated divergence time of chordates appears contemporaneous with or earlier than that of ambulacrarians. Together with paleontological data discussed in Chapter 3, this strongly suggests that the Ambulacraria and Chordata independently evolved. That is, chordates originated from a common ancestor shared with ambulacrarians but did not evolve downstream of ambulacrarians.

Chapter 5

Comparative Genomics of Deuterostomes

The genome contains all of the genetic information of a given organism; thus comparative genomics of deuterostome taxa is expected to offer insights into the origin and evolution of chordates. Now, decoded genomes of a sea urchin (Echinodermata), acorn worms (Hemichordata), lancelets (Cephalochordata), ascidians and a larvacean (Urochordata), and vertebrates (Vertebrata) are available. Comparisons of these genomes may tell us what genomic modifications were involved in diversification of ancestral deuterostomes into ambulacrarians, chordates, or vertebrates.

The genome contains all of the genetic information of a given organism, not only every developmentally relevant gene but also information about genomic organization, including synteny such as Hox clusters, intron-exon constitution, repetitive elements, transposable elements, etc. To understand biological phenomena of metazoans, including those associated with development, physiology, and evolution, decoding of all metazoan genomes is essential. If we obtain decoded genomes of all metazoan taxa, we may compare them to understand evolutionary scenarios of multicellular animals. Now, for the first time, decoded genomic information for all five deuterostome taxa was obtained in 2015. Comparisons of these genomes may allow us to identify genomic modifications that caused deuterostomes to diverge into ambulacrarians, chordates, and vertebrates. This chapter first provides basic information about the genome of each deuterostome phylum and discusses the most recent genomic advances in relation to chordate origins.

5.1 GENOME DECODING

In general, decoding the genome of a given organism is accomplished by four key processes: sequencing of fragmented genomic DNA (both nuclear and mitochondrial DNAs), assembly of sequenced fragments, annotation of genes within the assembly, and characterization of regulatory elements concealed in the genome. Transcriptome sequence data are essential for the last two steps. Sequencing technology has advanced dramatically in the last 10–15 years, with the development of second-generation (eg, Illumina) and third-generation

Chordate Origins and Evolution. http://dx.doi.org/10.1016/B978-0-12-802996-1.00005-5

sequencers (eg, PacBio), opening a new chapter in genome sciences. Bioinformatics tools to obtain better assembly, annotation, and gene modeling have also advanced with unexpected speed. Perhaps needless to say, hardware and software innovations have enhanced our understanding of the genomic basis of deuterostome evolution and chordate origins.

On the other hand, genomic data obtained thus far are not complete enough to discuss deuterostome evolution in the detail we desire. We face several difficulties in comparing decoded deuterostome genomes. First, because of comparatively high heterozygosity of specimens from wild populations, assemblies are sometimes not good enough for a comprehensive comparative analysis of the genomes. Second, most deuterostomes discussed in this book lack genetic backgrounds like those of *Drosophila* and *Caenorhabditis*, and this can create difficulties in discussion of, for example, genomic changes at the chromosomal level. Nevertheless, deuterostome genome decoding to date permits a reasonable discussion of chordate origins.

5.2 GENOMIC FEATURES OF FIVE REPRESENTATIVE DEUTEROSTOME TAXA

Table 5.1 briefly summarizes the composition and characteristic features of existing genomes in five deuterostome phyla.

5.2.1 Echinoderm Genomes

The draft genome of the sea urchin, *Strongylocentrotus purpuratus*, was decoded in 2006 (Sea Urchin Genome Sequence Consortium, 2006; Table 5.1). Because of the remarkable usefulness of its embryos as a model system for modern molecular developmental biology (eg, Davidson, 2006), it was logical that the sea urchin genome should be the first echinoderm genome decoded. The approximately 800 megabase-pair (Mbp) genome of *S. purpuratus* encodes approximately 23,300 protein-coding genes, including many previously thought to be vertebrate innovations or known only in protostomes. In addition to general information on gene families, it provided genomic backgrounds for several echinoderm-specific traits, such as the endoskeleton and the nonadaptive immune system.

Among deuterostomes, only echinoderms and vertebrates produce hard skeletons. The echinoderm skeleton consists of magnesium calcite, which is incorporated into many secreted matrix proteins, whereas vertebrates form calcium phosphate skeletons. The possible evolutionary relationship between biomineralization processes in these two groups has intrigued researchers for several decades. Analysis of the *S. purpuratus* genome revealed major differences in the proteins that mediate biomineralization in echinoderms and vertebrates (Livingston et al., 2006). First, there were few sea urchin counterparts of extracellular proteins that mediate biomineral deposition in gnathostome vertebrates. For example, secreted, calcium-binding phosphoproteins (SCPPs) are an important

TABLE 5.1 Comparative Genomics of Deuterostomes

Phylum	Species (Reference)	Genome Size (Mb)	Gene Number	Hox Cluster	Nkx2.1 Cluster	Features
Echinoderms						
Sea urchin	*Strongylocentrotus purpuratus*[a]	814	23,300	DO	NE	Exoskeleton formation-related genes
Star fish	*Acanthaster planci*[b]	440	26,000	O: Hox6 missing	O	Synteny of echinoderm genome
Hemichordates						
Acorn worm	*Saccoglossus kowalevskii*[c]	730	34,239	O	O	Deuterostome-related gene cluster
Acorn worm	*Ptychodera flava*[d]	980	34,687	O	O	Pharynx-related gene
Cephalochordates						
Lancelet	*Branchiostoma floridae*[e]	520	15,100	O: Hox14 and Hox15	O	Basic set of chordate genes
Lancelet	*Branchiostoma belcheri*[f]	420	30,400		NE	Conserved synteny with vertebrate genomes
Urochordates						
Ascidian	*Ciona intestinalis*[g]	155	15,600	DO	DO	Genome-wide reduction
Appendicularians	*Oikopleura dioica*[h]	64	29,600	DO	DO	Loss of synteny, CesA

Continued

TABLE 5.1 Comparative Genomics of Deuterostomes—cont'd

Phylum	Species (Reference)	Genome Size (Mb)	Gene Number	Hox Cluster	Nkx2.1 Cluster	Features
Vertebrates						
Hagfish	*Petromyzon marinus*[i]	816	26,500	2R-GWGD	NE	Cyclostome-specific genome constitution
Lamprey	*Lethenteron japonicum*[j]	1600		43 Hox genes	NE	
Shark	*Callorhinchus milii*[k]	940	18,880		NE	Slower evolutional rate of genes
Human	*Homo sapiens*[l]	3200	~25,000	4 paralogous groups	O	Pressure of repetitive segments

DO, disorganized; NE, not examined; 2R-GWGD, two rounds of genome-wide gene duplication.

[a]Sea Urchin Genome Sequence Consortium. 2006. *Science* 314, 941–952.
[b]Unpublished data.
[c]Simakov et al., 2015. *Nature* 527, 459–464.
[d]Simakov et al., 2015. *Nature* 527, 459–464.
[e]Putnam et al., 2008. *Nature* 453, 1064–1071.
[f]Huang et al., 2014. *Nature Communications* 5, 5896.
[g]Dehal et al., 2002. *Science* 298, 2157–2167.
[h]Denoeud et al., 2010. *Science* 330, 1381–1385.
[i]Smith et al., 2013. *Nature Genetics* 45, 415–421.
[j]Mehta et al., 2013. *Proceedings of the National Academy of Sciences of the United States of America* 110, 16044–16049.
[k]Venkatesh et al., 2014. *Nature* 505, 174–179.
[l]Intern. Human Genome Seq. Consortium 2001. *Nature* 409, 860–921.

class of proteins in vertebrate biomineralization. The sea urchin genome does not contain homologs of SCPP genes, suggesting that this family arose via a series of gene duplications after the echinoderm-chordate divergence. Second, almost all proteins directly implicated in control of biomineralization in sea urchins are specific to this clade. Sea urchin spicule matrix proteins are encoded by 16 genes organized in small clusters that likely proliferated by gene duplication. Given an early divergence of ambulacrarian and chordate lineages, it is evident that biomineralization has developed independently in echinoderms and vertebrates.

The *S. purpuratus* genome shows that echinoderms have a greatly expanded repertoire of innate immune system proteins compared with any other animal studied to date. Three classes of innate receptor proteins that are particularly expanded include a large family of Toll-like receptors, an equally large family of genes encoding NACHT and leucine-rich repeat-containing proteins, and a set of genes encoding multiple scavenger receptor cysteine-rich domain proteins. This last class is highly expressed in sea urchin immune cells or coelomocytes. In contrast, urchin homologs of signal transduction proteins and nuclear factor-kappa B/Rel domain transcription factors that are known to function further downstream of these genes were present in numbers similar to those in other invertebrate species.

Further genome-decoding projects relative to other echinoderm classes are in progress (Cameron et al., 2015). A question frequently encountered when we discuss the evolution of metazoans is "which species can legitimately represent its phylum?" For example, sea urchin development, which involves micromere formation at the 16-cell stage, giving rise to the larval skeleton, seems derived in echinoderms. In this respect, the genome of the crown-of-thorns starfish, *Acanthaster planci* (a starfish expanding its population in the Indo-Pacific ocean and increasingly problematic in coral reef management because it eats corals), appears to provide more useful information for comparison of ambulacrarian and chordate genomes. The *A. planci* genome is approximately 430 Mbp in size and contains approximately 24,500 protein-coding genes. Less heterozygosity (0.9%) has allowed the best assembly of an echinoderm genome thus far, which facilitates comparison of synteny, as discussed in the section on Hox clusters (Section 6.3 and Baughman et al., 2014) and Nkx2.1 cluster (Section 6.2 and Simakov et al., 2015).

5.2.2 Hemichordate Genomes

As discussed in Section 1.3.2, enteropneusts, or "gut-breathing acorn worms," a major hemichordate group, share features with echinoderms and chordates. Acorn worm genomes may give us insights into shared traits that were likely inherited from the last common deuterostome ancestor and may permit us to explore evolutionary trajectories leading from this ancestor to hemichordates and chordates. Indeed, the draft genomes of two acorn worms, *Saccoglossus kowalevskii* (a direct developer in which tornaria and Müller larval stages are skipped) and *Ptychodera flava* (an indirect developer that shows a comparatively long larval stage before metamorphosis), have provided us at least three novel findings in relation to deuterostome evolution (Simakov et al., 2015). Both genomes have haploid lengths of approximately 1 gigabase-pair (Gbp; Table 5.1), but they differ in nucleotide heterozygosity (0.5% in *S. kowalevskii* and 1.3% in *P. flava*). *Ptychodera* and *Saccoglossus* encode at least 18,556 and 19,270 genes, respectively. Despite the ancient divergence of their lineages [>370 million years ago (MYA); Fig. 4.2] and their different modes of development previously mentioned, the two acorn worm genomes have similar bulk gene content and similar repetitive landscapes.

First, as discussed in Chapter 4, molecular phylogenomics in the acorn worm genome project, using not only amino acid characters but also presence-absence characters for introns and coding indels, unambiguously demonstrated the monophyly of hemichordates, including pterobranchs, ambulacrarians, chordates, and deuterostomes (Fig. 4.1). Second, hemichordate genomes exhibit extensive conserved synteny with cephalochordates (Fig. 5.1A). Their Hox cluster is comparable to that of cephalochordates, standard among deuterostomes (Section 6.3). In addition, they strikingly possess a deuterostome-specific genomic cluster of four ordered transcription factor genes, expression

FIGURE 5.1   Genome-wide synteny conservation among deuterostomes. (A) A high level of linkage conservation between the hemichordate, *Saccoglossus kowalevskii*, and the cephalochordate, *Branchiostoma floridae*, shown by macro-synteny dot plots. Each dot represents two orthologous genes linked in the two species, and ordered according to their macro-syntenic linkage. Intersection areas of highest dot density are marked, identifying each of the 17 putative ancestral linkage groups. *(From Simakov, O., Kawashima, T., Marlétaz, F., Jenkins, J., Koyanagi, R., Mitros, T., Hisata, K., Bredeson, J., Shoguchi, E., Gyoja, F., Yue, J.X., Chen, Y.C., Freeman, R.M., Sasaki, A., Hikosaka-Katayama, T., Sato, A., Fujie, M., Baughman, K.W., Levine, J., Gonzalez, P., Cameron, C., Fritzenwanker, J.H., Pani, A.M., Goto, H., Kanda, M., Arakaki, N., Yamasaki, S., Qu, J., Cree, A., Ding, Y., Dinh, H. H., Dugan, S., Holder, M., Jhangiani, S.N., Kovar, C.L., Lee, S.L., Lewis, L. R., Morton, D., Nazareth, L.V., Okwuonu, G., Santibanez, J., Chen, R., Richards, S., Muzny, D.M., Gillis, A., Peshkin, L., Wu, M., Humphreys, T., Su, Y.H., Putnam, N.H., Schmutz, J., Fujiyama, A., Yu, J.K., Tagawa, K., Worley, K.C., Gibbs, R.A., Kirschner, M.W., Lowe, C.J., Satoh, N., Rokhsar, D.S., Gerhart, J., 2015. Hemichordate genomes and deuterostome origins. Nature 527, 459–465.)* (B) Quadruple, conserved synteny between (a) amphioxus and (b) vertebrate genomes. Partitioning of human chromosomes into segments with defined patterns of conserved synteny (b) to amphioxus (*B. floridae*) scaffolds (a). Numbers 1–17 in (a) represent 17 reconstructed ancestral chordate linkage groups, and letters a–d represent four products resulting from two rounds of genome-wide gene duplication. Colored bars in (b) are segments of the human genome, grouped by ancestral linkage group (above), and in the context of the human chromosomes. *(From Putnam, N.H., Butts, T., Ferrier, D.E.K., Furlong, R.F., Hellsten, U., Kawashima, T., Robinson-Rechavi, M., Shoguchi, E., Terry, A., Yu, J.K., Benito-Gutiérrez, E., Dubchak, I., Garcia-Fernàndez, J., Gibson-Brown, J.J., Grigoriev, I.V., Horton, A.C., de Jong, P.J., Jurka, J., Kapitonov, V.V., Kohara, Y., Kuroki, Y., Lindquist, E., Lucas, S., Osoegawa, K., Pennacchio, L.A., Salamov, A.A., Satou, Y., Sauka-Spengler, T., Schmutz, J., Shin-I, T., Toyoda, A., Bronner-Fraser, M., Fujiyama, A., Holland, L.Z., Holland, P.W.H., Satoh, N., Rokhsar, D.S., 2008. The amphioxus genome and the evolution of the chordate karyotype. Nature 453, 1064–1071.)*

(A)

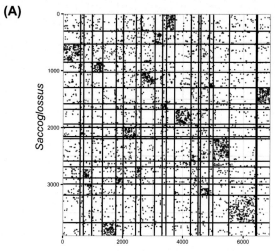

(B)

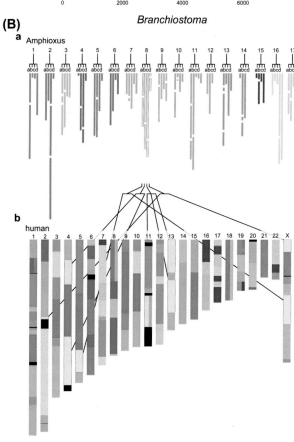

of which is associated with development of the gill slit-containing pharyngeal apparatus, the foremost morphological innovation of deuterostomes that is central to their filter-feeding lifestyle. This *Nkx2.1-Nkx2.2-Pax1/9-FoxA* cluster will be discussed in Section 6.2. Third, comparative analysis also revealed numerous deuterostome-specific gene novelties, including genes found in deuterostomes and marine microbes but not other animals (Chapter 11). These novelties can be linked to physiological, metabolic, and developmental specializations of the filter-feeding ancestor. Further deep analyses of hemichordate genomes are now ongoing.

5.2.3 Cephalochordate Genome

Lancelet genomes retain basic components of chordate genomes, as was expected. The 520-Mbp-long genome of the Florida lancelet, *Branchiostoma floridae*, contains approximately 21,900 protein-coding loci (Putnam et al., 2008; Table 5.1). A recently decoded genome of the Chinese lancelet, *Branchiostoma belcheri*, also has gene content very similar to that of *B. floridae* (Huang et al., 2014). *B. floridae* gene families have been intensively examined, with special attention given to homeobox genes, opsin genes, genes involved in neural crest development, nuclear receptor genes, genes encoding components of the endocrine and immune systems, and conserved *cis*-regulatory enhancers (Holland et al., 2008). The amphioxus genome contains a basic set of chordate transcription factor genes and signal pathway molecule genes, whereas Olfactores, especially urochordates, have lost many of these (Fig. 5.2), including various homeobox genes and those involved in steroid hormone function, as discussed in Section 6.3.

The most striking finding of the *B. floridae* genome is extensive conservation of synteny between cephalochordates and vertebrates (Putnam et al., 2008). The pattern of amphioxus-human synteny, revealed by reciprocal cluster analysis of both amphioxus scaffolds and human chromosomal segments, has identified 17 ancient chordate linkage groups (CLGs; Fig. 5.1B). Using fluorescent in situ hybridization (FISH), these CLGs were confirmed to be coherently evolving segments. In addition, analysis to obtain a genome-wide view of chromosomal evolution along the vertebrate stem showed that most of the human genome was affected by large-scale duplication events on the vertebrate stem, and that nearly all of the ancient chordate chromosomes were quadruplicated (Fig. 5.1B). This pattern of genome-wide quadruple conserved synteny definitively confirms the occurrence of two rounds of genome-wide gene duplication (2R-GWGD) and provides a comprehensive reconstruction of the evolutionary origin of vertebrate chromosomes (see Fig. 5.1B). This characterization extends previous lines of evidence for 2R-GWGD events based on comparative studies of specific regions of interest across chordate genomes (eg, the Hox cluster, myogenic regulatory genes, and the major histocompatibility complex region) and the analysis of vertebrate gene families (Panopoulou et al., 2003) as well

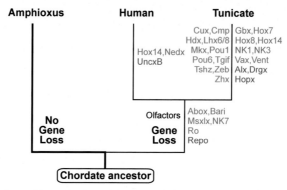

FIGURE 5.2 Loss of homeobox-containing genes during the evolution of tunicates. Homeobox gene loss has been extensive during evolution of the Olfactores (vertebrates plus tunicates) and tunicates since the last common ancestor of chordates. (*Blue*) ANTP (antennapedia) class; (*red*) PRD (paired) class; (*green*) other classes. Gene losses along the tunicate branch occurred before the last common ancestor of *Ciona* and *Oikopleura*. (*From Holland, L.Z., Albalat, R., Azumi, K., Benito-Gutiérrez, E., Blow, M.J., Bronner-Fraser, M., Brunet, F., Butts, T., Candiani, S., Dishaw, L.J., Ferrier, D.E., Garcia-Fernàndez, J., Gibson-Brown, J.J., Gissi, C., Godzik, A., Hallböök, F., Hirose, D., Hosomichi, K., Ikuta, T., Inoko, H., Kasahara, M., Kasamatsu, J., Kawashima, T., Kimura, A., Kobayashi, M., Kozmik, Z., Kubokawa, K., Laudet, V., Litman, G.W., McHardy, A.C., Meulemans, D., Nonaka, M., Olinski, R.P., Pancer, Z., Pennacchio, L.A., Pestarino, M., Rast, J.P., Rigoutsos, I., Robinson-Rechavi, M., Roch, G., Saiga, H., Sasakura, Y., Satake, M., Satou, Y., Schubert, M., Sherwood, N., Shiina, T., Takatori, N., Tello, J., Vopalensky, P., Wada, S., Xu, A., Ye, Y., Yoshida, K., Yoshizaki, F., Yu, J.K., Zhang, Q., Zmasek, C.M., de Jong, P.J., Osoegawa, K., Putnam, N.H., Rokhsar, D.S., Satoh, N., Holland, P.W., 2008. The amphioxus genome illuminates vertebrate origins and cephalochordate biology. Genome Research 18, 1100–1111.*)

as the identification of paralogous segments and chromosomal relationships within the human genome. Together with other gene family evidence, cephalochordates appear to have retained the genomes of both ancient chordates and ancient deuterostomes.

5.2.4 Urochordate Genomes

As previously mentioned, cephalochordates offer a reference deuterostome genome and have well-conserved synteny to vertebrates. In addition, genomes of animals as different as the cnidarian, *Nematostella vectensis*, and humans show conservation of global synteny architecture (Putnam et al., 2007). In contrast, certain animals with very short lifecycles or parasitic lifestyles show a dramatic change in their genome compositions. Tunicates show remarkable changes in genomic constitution, which are probably associated with their developmental mode and filter-feeding lifestyle. Specifically, urochordate genomes have been compacted (Table 5.1). For example, the genome size of *Ciona intestinalis* is 169 Mbp. This trend is remarkable in the case of larvaceans; the *Oikopleura dioica* genome is only 70 Mbp, with a generation time of only approximately 4 days, at 20°C (eg, Nishino and Satoh, 2001; Nishida, 2008).

5.2.4.1 Ciona intestinalis

The *C. intestinalis* genome contains approximately 15,600 protein-coding genes (Satou et al., 2008). Vertebrate gene families are typically found in simplified form in *Ciona*, suggesting that ascidians contain the basic ancestral complement of developmentally relevant genes. The ascidian genome has also acquired several lineage-specific innovations, including a group of genes engaged in cellulose metabolism, obtained by horizontal gene transfer from bacteria (Chapter 11). Genome comparisons among *B. floridae*, *Homo sapiens*, and *C. intestinalis* suggests that the last common ancestor of chordates possessed 103 homeobox genes and that 6, 3, and 21 genes were lost during evolution of the Olfactores, vertebrates, and tunicates, respectively (Fig. 5.2; Holland et al., 2008). In contrast to the highly conserved synteny between cephalochordates and vertebrates, very few synteny blocks were found between *C. intestinalis* and vertebrates. In addition, as discussed later, the Hox cluster (Section 6.3) and the Nkx2.1 cluster (Section 6.2) of *C. intestinalis* became disorganized.

On the other hand, genes of the *C. intestinalis* genome, especially those encoding transcription factors and signal pathway molecules, have been annotated most intensively among deuterostome genomes (eg, Satoh, 2014). In addition, genome sequences have been mapped on 14 pairs of chromosomes using an elaborate FISH strategy (Shoguchi et al., 2008). Together with its simple mode of embryogenesis, the *C. intestinalis* genome provides a basis for studies of genome-wide gene regulatory networks involved in the development of basic chordate body plans (Imai et al., 2006; Satoh, 2014; Brozovic et al., 2016).

5.2.4.2 Oikopleura dioica

The *O. dioica* genome is only 70 Mbp in size (Table 5.1; Denoeud et al., 2010). Introns are very small (≤47 Mbp), as are intergenic spaces, partly because of numerous operons. Genes outside operons are also densely packed, and the density of transposable elements is low. In addition, there has been a significant gene loss in the urochordate genome. However, this genome suggests some traits of sex chromosome development.

Altogether, urochordates have very different genomes from those of other deuterostomes. This genomic feature likely provides the basis for modifications of embryogenesis and adaptation of various classes to their lifestyles.

5.2.5 Vertebrate Genomes

Vertebrates comprise jawless vertebrates (lampreys and hagfishes), cartilaginous vertebrates, and bony vertebrates. The origin and evolution of vertebrates from a common chordate ancestor rely on evo-devo events too numerous to discuss in this book. Various vertebrate genomes are targets of recent global sequencing projects, such as the 10k project (https://genome10k.soe.ucsc.edu/). Here, in the context of the early phase of vertebrate evolution, genomes of lampreys, elephant sharks, and puffer fish are briefly described.

5.2.5.1 Lampreys

Lampreys are representatives of an ancient vertebrate lineage that diverged from gnathostome vertebrates approximately 485 MYA. A striking phenomenon of the lamprey genome is that the lamprey undergoes programmed genome rearrangement during embryogenesis, which results in deletion of approximately 20% of germ-line DNA from somatic tissues, although the effects of this rearrangement on the genic component of the genome are not fully understood (Smith et al., 2009). In addition, repetitive elements that occupy approximately 35% of the genome and high GC content in protein-coding regions (61%) make it difficult to appropriately assemble the DNA sequence. Nevertheless, the genome of the sea lamprey, *Petromyzon marinus*, is approximately 800 Mbp in size and contains approximately 25,000 protein-coding genes (Table 5.1; Smith et al., 2013). The *P. marinus* genome resolved the long-standing question of whether 2R-GWGD occurred in vertebrate evolution (Section 5.2.3). Analyses of the *P. marinus* assembly indicate that it likely occurred before the divergence of the ancestral lamprey and gnathostome lineages. This conclusion is supported by the draft genome assembly of another lamprey, *Lethenteron japonicum* (Table 5.1; Mehta et al., 2013). In addition, the lamprey genomes help to define key evolutionary events within vertebrate lineages, including the origin of myelin-associated proteins and the development of appendages.

5.2.5.2 Elephant Shark

The emergence of jawed vertebrates (gnathostomes) from jawless vertebrates was accompanied by major morphological and physiological innovations, such as hinged jaws, paired fins, and immunoglobulin-based adaptive immunity. Gnathostomes subsequently diverged into two groups: cartilaginous fishes and bony vertebrates. A decoded genome of a cartilaginous fish, the elephant shark, *Callorhinchus milii*, provides evolutionary insight into early jawed vertebrates (Table 5.1; Venkatesh et al., 2014).

The genome of *C. milii* is approximately 940 Mbp in size and contains approximately 18,900 protein-coding genes. It is interesting to note that the *Callorhinchus* genome is the slowest evolving of all known vertebrates, including the coelacanth, regarded as a "living fossil." There are features of extensive synteny conservation with tetrapod genomes, making it a good model for comparative analyses of gnathostome genomes. In addition, functional analyses suggest that the lack of genes encoding SCPPs in cartilaginous fishes explains the absence of bone in their endoskeleton. The *Callorhinchus* genome contains *sparc-like 1* (*Sparcl1*), which is highly likely to have duplicated tandemly in the Osteichthyes, one remaining as the original *Sparcl1* and the other becoming *SCPP*. In addition, it was shown that the adaptive immune system of cartilaginous fishes is unusual. It lacks the canonical CD4 co-receptor and most transcription factors, cytokines, and cytokine receptors related to the CD4 lineage despite the presence of polymorphic, major histocompatibility complex class II molecules. Thus *Callorhinchus* provides a model for understanding the origin of adaptive immunity (Venkatesh et al., 2014).

5.2.5.3 Fish

The compact genome of *Fugu rubripes* was sequenced to greater than 95% coverage, and more than 80% of the assembly was in multigene-sized scaffolds (Aparicio et al., 2002). In this 365-Mbp vertebrate genome, repetitive DNA accounts for less than one-sixth of the genome, and gene loci occupy approximately one-third of it. As with the human genome, gene loci are not evenly distributed but are clustered into sparse and dense regions. Some "giant" genes were observed that had average coding sequence sizes but were spread over genomic lengths significantly larger than those of their human orthologs. Although three-quarters of predicted human proteins match those of *Fugu*, approximately one-quarter of the human proteins had highly diverged from or had no pufferfish homologs, highlighting the extent of protein evolution in approximately 360 million years (MY) since teleosts and mammals diverged. Conserved linkages between *Fugu* and human genes indicate the preservation of chromosomal segments from the common vertebrate ancestor but with considerable scrambling of gene order.

5.3 GENE FAMILIES IN DEUTEROSTOMES AND THE ANCESTRAL GENE SET

As previously described, genomic comparisons of five deuterostome taxa illustrate several trends. First, although echinoderms, hemichordates, and cephalochordates possess genomes of comparable size, that of urochordates is reduced, presumably because of their short generation times, whereas the vertebrate genome is expanded because of 2R-GWGD (Ohno, 1970; Holland et al., 1994). Second, genomes of hemichordates, cephalochordates, and vertebrates share a high level of synteny (Fig. 5.1), but urochordates lack it. Although further analysis is required for echinoderm genomes, they likely conserve a level of synteny. Third, several genes have emerged in specific lineages.

Putnam et al. (2007) categorized gene family novelty into four types (type IV is the ancestral form shared by all eumetazoans). Type I novelty comprises animal genes that have no relatives identifiable with Basic Local Alignment Search Tool (BLAST) beyond the species in the available genomes (see also Section 5.8). Type I accounts for approximately 15% of ancestral metazoan genes. These include important signal pathway molecules, such as the secreted wingless (Wnt) and fibroblast growth factor (FGF) families, and transcription factors, including the mothers-against-decapentaplegic (SMAD) family. Not only were these genes present in the eumetazoan ancestor, but they had already duplicated and diversified on the eumetazoan stem to establish subfamilies that are still maintained in modern vertebrates nearly 570 MY later. Type II novelties incorporate animal-only domains inserted into ancient eukaryotic sequences. Ancestry of these genes can be traced back to the eukaryotic radiation through their ancient domains, but the novel domains were evidently invented (or evolved into their present forms) and fused with more ancient domains on the eumetazoan stem. For example, Notch proteins have two Notch domains found only in metazoans in addition to ancient eukaryotic ankyrin and epidermal growth factor (EGF) domains. Type III novelties consist of animal genes containing only ancient domains (ie, found in other eukaryotes), but that occur in combinations apparently unique to eumetazoans, because of gene fusions and/or domain-shuffling events on the eumetazoan stem. For example, both the LIM (lin-11, islet, mec-3) protein–protein interaction and homeobox DNA-binding domains are found in nonanimal eukaryotes, but only animals have the LIM-homeodomain combination. This categorization is useful for further discussions of molecular evolution relative to chordate origin and evolution.

5.3.1 Gene Families in Deuterostomes

How many gene families are shared by all deuterostome taxa? What kinds of qualitative and quantitative changes have occurred in these families during deuterostome evolution (eg, Simakov et al., 2013)? How did these changes cause deuterostome, ambulacrarian, or chordate novelties? Custom clustering analysis shows that deuterostomes, ambulacrarians, and chordates each share 8716, 9892, and 9957 gene families, respectively (Table 5.2), implying the presence of at least 8716 families of homologous genes in the deuterostome ancestor (Simakov et al., 2015). Because of gene duplication and other processes, descendants of these ancestral genes account for approximately 14,000 genes in extant deuterostome genomes, including that of humans. *Saccoglossus*, *Branchiostoma*, and *Homo* now possess 7256, 7209, and 6419, respectively, of the 8716 ancestral deuterostome gene families (Table 5.2).

Table 5.2 shows that deuterostomes, ambulacrarians, and chordates have 369, 425, and 438 novel gene families, respectively. Further analyses demonstrate that nearly 30 deuterostome-specific genes introduce possible functional innovations. Some plausibly arose from accelerated sequence changes on the deuterostome stem from distant, but identifiable bilaterian homologs. Others represent new protein domain combinations in deuterostomes, while others lack identifiable sequence and domain homologs in other animals. In the latter group, there are over a dozen deuterostome genes that have readily identifiable homologs in marine microbes, often cyanobacteria or eukaryotic micro-algae, but that are not known in other metazoans. Such genes include two of the novel deuterostome sequences associated with sialic acid metabolism (found in many microbes), details of which are discussed in Chapter 11. Another novelty found in deuterostomes is discussed in Chapter 6.

5.3.2 Expansion of Gene Families in Deuterostomes

As previously discussed, the vertebrate genome has experienced both quantitative and qualitative alterations as a result of 2R-GWGD that occurred in the early phase of vertebrate divergence. Indeed, numerous gene families, including

TABLE 5.2 Gene Novelties and Losses of Genes in the Metazoans and Deuterostomes

Node	Bilaterians	Deuterostomes	Ambulacrarians	Chordates
Criteria	At least 2 species in protostomes, deuterostomes, or outgroup	At least 2 species in ambulacrarians, chordates, or outgroup	Sea urchin and at least 1 hemichordate or 2 in the outgroup	At least 1 in Amphioxus and Ciona, and 2 in vertebrates, or 2 in the outgroup
MRCA (most recent common ancestor) gene families	8423	8716	9892	9957
Present in Sko	6832	7256	8392	737
Present in Pfl	5179	5534	6361	5446
Present in Spu	6441	6757	7866	6889
Present in Bfl	6992	7209	7385	8398
Novelties	N/A	369	425	438
Type 1 novelties	N/A	20	27	27
Losses	N/A	N/A	3074	1207
Type 1 losses	N/A	N/A	183	156

Sko, Saccoglossus kowalevskii; Pfl, Ptychodera flava; Spu, Strongylocentrotus purpuratus; Bfl, Branchiostoma floridae.
Novelty, following Putnam et al., 2007. Science 317, 86–94; type 1 indicates gains or losses with no similarity to outgroups or ingroups, respectively. Courtesy of Oleg Simakov.

those encoding transcription factors (Hox, ParaHox, En, Otx, Msx, Pax, Dlx, HNF3, bHLH, etc.), signal pathway molecules (hh, IGF, BMP, etc.), and various other proteins (dystrophin, cholinesterase, actin, keratin, etc.), were expanded by gene duplication in the vertebrate stem lineage. This resulted in increased genetic and morphological complexity, along with developmental control. On the other hand, generally speaking, what differences exist in gene families of decoded metazoan genomes?

Results of two analyses to assess broad trends in gene family expansion and/ or loss are shown in Fig. 5.3. Fig. 5.3A shows clustering of metazoan genomes in a multidimensional space of molecular functions using Panther annotations that identify patterns in gene family representation across metazoans. In this analysis, the first principal component (PC) (12% of variance) distinguishes vertebrates from other metazoans, and PC2 (7%) distinguishes ecdysozoans from the rest, including diploblasts, spiralians, and nonvertebrate deuterostomes. Similar results are obtained with Protein Family data base (PFAM) annotations. Fig. 5.3B shows the result of gene family comparison among metazoans, including the genome of the brachiopod, *Lingula anatina* (Luo et al., 2015). This study exploited gene families of 22 taxa with decoded metazoan genomes. Heatmap analysis clearly shows that (1) the content of gene families in vertebrate genomes is distinguishable from invertebrate genomes; (2) there are no conspicuous differences among nonvertebrate deuterostomes, including tunicates, cephalochordates, and echinoderms; (3) the content of gene families in nonvertebrate deuterostomes overlaps with those of lophotrochozoans; and (4) the family content of ecdysozoans is different from those of other bilaterians. Altogether, the gene family repertoire of vertebrates is quite different because of whole-genome duplication and moderately different in ecdysozoans as well. In contrast, the same gene families are comparatively common among all nonvertebrate deuterostomes. The distribution of gene functions, domain compositions, and gene family sizes is similar among echinoderms, hemichordates, cephalochordates, urochordates, and even lophotrochozoans. Therefore it is safe to say that the origin and evolution of nonvertebrate chordates occurred in gene families shared with ambulacrarians.

5.4 EXON-INTRON STRUCTURES

Most eukaryote genes are composed of exons and introns. Exon-intron structure (ie, how many introns subdivide a given gene) and where exons are spliced by introns have been discussed frequently in relation to evolutionary modification of gene function. Comparative genomics shows that exon-intron structures are generally well conserved among bilaterians and deuterostomes (Simakov et al., 2015). Approximately 2000 splice sites are likely to be conserved among many nondeuterostome metazoans, hemichordates, and chordates, suggesting ancestral deuterostome splice sites. Analyses of exon-intron structures among orthologous bilaterian genes show that 23 introns and 4 coding sequence indels are present only in deuterostomes (shared by at least one ambulacrarian and

(A)

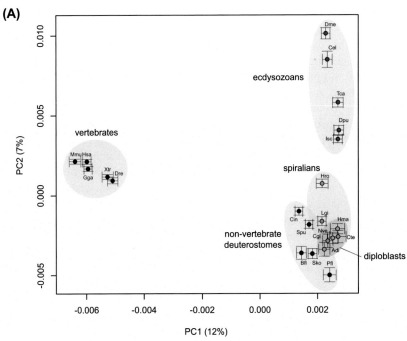

(B)

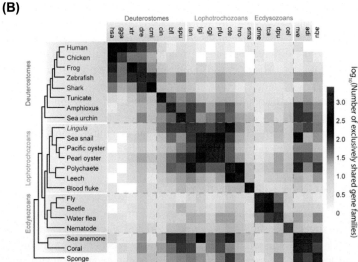

FIGURE 5.3 Gene family expansions in metazoans. (A) Clustering of metazoan genomes in a multidimensional space of molecular functions. Clustering was done using Panther annotations that identify common patterns in gene family representation among metazoans. These annotations represent each genome as a point in a high-dimensional "function space," where coordinates are the number of genes in each functional category (eg, Protein Family data base (PFAM) domain). For this Panther-based PCA (principal component analysis), the first PC (principal component) (12% of variance) distinguishes vertebrates from other metazoans, and PC2 (7%) distinguishes ecdysozoans from the

one chordate). These shared and derived characters may be useful to diagnose membership of new candidate clades such as the Xenacoelomorpha. On the other hand, in general, it may be difficult to infer novelty of gene function from exon-intron structure.

5.5 SYNTENY

Synteny is a conserved similarity of gene order in a given genomic or chromosomal region. Sometimes it extends to the chromosome level (macrosynteny or ancient gene linkages) or is restricted to a small genomic region (microsynteny). Macrosynteny indicates conservation of genomic constitution or gene content that suggests evolutionary intimacy of two phyla. Microsynteny indicates gene clusters that have conserved functional roles in development and/or physiology.

5.5.1 Macrosynteny

It has been shown that some degree of macrosynteny is evident between genomes of the cnidarian *Nematostella vectensis* and vertebrates (Putnam et al., 2007), suggesting the evolutionary origin of metazoans. As discussed in Section 5.2.3, the cephalochordate, *B. floridae*, and humans exhibit considerable macrosynteny, making it possible to infer an ancestral chordate karyotype (Fig. 5.1B; Putnam et al., 2008). Macrosynteny is also seen between hemichordate

rest. *Blue*, deuterostomes; *green*, spiralians; *red*, ecdysozoans; *yellow*, nonbilaterian metazoans. Error bars represent variation based on 100 dataset randomizations with replacement. At least three clusters are evident, including a vertebrate cluster (*red circle*); a nonbilaterian metazoan, invertebrate deuterostome, or spiralian cluster (*green circle*); and an ecdysozoan group (*yellow circle*). *Aqu, Amphimedon queenslandica* (demosponge); *Bfl, Branchiostoma floridae* (amphioxus); *Cel, Caenorhabditis elegans*; *Cin, Ciona intestinalis* (sea squirt); *Cte, Capitella teleta* (polychaete); *Dme, Drosophila melanogaster*; *Dpu, Daphnia pulex* (water flea); *Dre, Danio rerio* (zebrafish); *Gga, Gallus gallus* (chicken); *Hma, Hydra magnipapillata*; *Hro, Helobdella robusta* (leech); *Hsa, Homo sapiens* (human); *Isc, Ixodes scapularis* (tick); *Lgi, Lottia gigantea* (limpet); *Mmu, Mus musculus* (mouse); *Nve, Nematostella vectensis* (sea anemone); *Pfl, Ptychodera flava*; *Sko, Saccoglossus kowalevskii*; *Sma, Schistosoma mansoni*; *Sme, Schmidtea mediterranea* (planarian); *Spu, Strongylocentrotus purpuratus* (sea urchin); *Tad, Trichoplax adhaerens* (placozoan); *Tca, Tribolium castaneum* (flour beetle); *Xtr, Xenopus tropicalis* (clawed frog). *(From Simakov, O., Kawashima, T., Marlétaz, F., Jenkins, J., Koyanagi, R., Mitros, T., Hisata, K., Bredeson, J., Shoguchi, E., Gyoja, F., Yue, J.X., Chen, Y.C., Freeman, R.M., Sasaki, A., Hikosaka-Katayama, T., Sato, A., Fujie, M., Baughman, K.W., Levine, J., Gonzalez, P., Cameron, C., Fritzenwanker, J.H., Pani, A.M., Goto, H., Kanda, M., Arakaki, N., Yamasaki, S., Qu, J., Cree, A., Ding, Y., Dinh, H. H., Dugan, S., Holder, M., Jhangiani, S.N., Kovar, C.L., Lee, S.L., Lewis, L. R., Morton, D., Nazareth, L.V., Okwuonu, G., Santibanez, J., Chen, R., Richards, S., Muzny, D.M., Gillis, A., Peshkin, L., Wu, M., Humphreys, T., Su, Y.H., Putnam, N.H., Schmutz, J., Fujiyama, A., Yu, J.K., Tagawa, K., Worley, K.C., Gibbs, R.A., Kirschner, M.W., Lowe, C.J., Satoh, N., Rokhsar, D.S., Gerhart, J., 2015. Hemichordate genomes and deuterostome origins. Nature 527, 459–465.)* (B) Heatmap of exclusively shared gene families among metazoans. Higher values indicate higher similarity. This analysis also supports at lease for clusters, including a vertebrate cluster (left corner); a nonbilaterian metazoan, invertebrate deuterostome, or spiralian cluster (middle); and an ecdysozoan group (lower right) and diploblasts cluster (right corner). *(Courtesy of Yi-Jyun Luo.)*

and cephalochordate genomes (Fig. 5.1A). In contrast, almost no sign of macrosynteny is found between cephalochordate and urochordate genomes because of the highly derived states of ascidians and larvaceans. It is tempting to ask whether macrosynteny exists between echinoderms and cephalochordates. Nevertheless, the extent of macrosynteny conservation among deuterostomes suggests that hemichordates-cephalochordates-vertebrates comprise the main stem leading to vertebrates whereas echinoderms and urochordates diverged, as evidenced by the acquisition of specific characters, a mineralized exoskeleton in echinoderms, and a cellulosic tunic in urochordates.

5.5.2 Microsynteny

Microsynteny is better characterized than macrosynteny because conservation of microsyntenic linkages occurs more readily because of low rates of localized genomic rearrangement, or more importantly because of selection that retains linkages between genes and their regulatory elements, located in neighboring genes. Because of its small size, microsynteny is easy to identify among genomes. There are hundreds of tightly linked, conserved clusters of three or more genes. Hox and ParaHox clusters are one of the most intensively examined microsyntenies: these are discussed in Section 6.3 in relation to chordate origins and evolution. The Wnt gene family is another example. In addition, the hemichordate genome project revealed the presence of a microsynteny of four transcription factors.

Fig. 5.4 shows four examples of microsynteny conserved in deuterostome genomes (Simakov et al., 2015). For example, *Univin* is tightly linked to the related bilaterian *bmp2/4* in genomes of the sea urchin, hemichordates, and amphioxus, supporting its origin by tandem duplication and divergence from an ancestral *bmp2/4*-type gene, as previously suggested (Fig. 5.4A; Simakov et al., 2015) and discussed further in Chapter 11.

5.6 CONSERVED NONCODING SEQUENCES

Vertebrate genome comparisons, especially genomes of several mammals, have demonstrated the presence of 50-bp or more, highly conserved nucleotide sequences in intergenic or intron regions of the genomes. Although these conserved noncoding sequences (CNSs) or elements (CNEs) have a tendency to get lost when genomes are compared between more distantly related taxa, they are believed to contain *cis*-regulatory modules, essential to gene expression and to other transcriptionally important sequences (eg, Hufton et al., 2009; Sanges et al., 2013).

The hemichordate genome project has identified 6533 CNEs longer than 50 bp that are found in all five deuterostome genomes, *Saccoglossus, Ptychodera, Branchiostoma, Strongylocentrotus,* and *Homo* (Simakov et al., 2015). Identified CNEs overlapped extensively with human long-noncoding RNAs (3611 CNE loci). Those overlap alignments usually do not exceed 250 bp (as has been reported among vertebrates), and they occur in clusters. One CNS

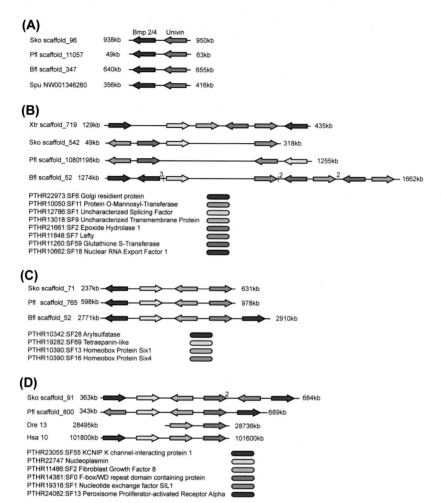

FIGURE 5.4 Deuterostome-specific microsyntenic linkages. (A) *bmp2/4* and *univin* cluster in the hemichordates *S. kowalevskii* (Sko) and *P. flava* (Pfl), the sea urchin *S. purpuratus* (Spu), and the cephalochordate *B. floridae* (Bfl). This linkage is tight, with no intervening genes. (B–D) Loose microsyntenic linkages with a maximum of five intervening genes: *lefty* (B), *six1–six4* (C), and *fgf8–fbxw* (D) clusters. *Xtr, Xenopus tropicalis; Hsa, Homo sapiens. (From Simakov, O., Kawashima, T., Marlétaz, F., Jenkins, J., Koyanagi, R., Mitros, T., Hisata, K., Bredeson, J., Shoguchi, E., Gyoja, F., Yue, J.X., Chen, Y.C., Freeman, R.M., Sasaki, A., Hikosaka-Katayama, T., Sato, A., Fujie, M., Baughman, K.W., Levine, J., Gonzalez, P., Cameron, C., Fritzenwanker, J.H., Pani, A.M., Goto, H., Kanda, M., Arakaki, N., Yamasaki, S., Qu, J., Cree, A., Ding, Y., Dinh, H. H., Dugan, S., Holder, M., Jhangiani, S.N., Kovar, C.L., Lee, S.L., Lewis, L. R., Morton, D., Nazareth, L.V., Okwuonu, G., Santibanez, J., Chen, R., Richards, S., Muzny, D.M., Gillis, A., Peshkin, L., Wu, M., Humphreys, T., Su, Y.H., Putnam, N.H., Schmutz, J., Fujiyama, A., Yu, J.K., Tagawa, K., Worley, K.C., Gibbs, R.A., Kirschner, M.W., Lowe, C.J., Satoh, N., Rokhsar, D.S., Gerhart, J., 2015. Hemichordate genomes and deuterostome origins. Nature 527, 459–465.)*

overlaps an experimentally verified enhancer (element 488, specific for vertebrate brain and neural tube), located close to the *sox14/21* ortholog in all five species. Deep analyses may identify CNEs and their modified forms that are associated with novel and/or modified modes of genes, especially those encoding transcription factors or signal pathway molecules.

5.7 REPETITIVE ELEMENTS

More than 40% of the human genome consists of repetitive sequences, including DNA transposons, long terminal repeat (LTR) retrotransposons, non-LTR retrotransposons, etc. (eg, Brown, 2002). Quality and quantity of repetitive elements are roughly similar across deuterostomes (and even other metazoans: http://www.repeatmasker.org/genomicDatasets/RMGenomicDatasets.html).

For example, the total repetitive element content in the acorn worm, *S. kowalevskii,* is 33.3% (Simakov et al., 2015). Simple or low-complexity repeats are the most common class (5–7% of the genome), followed by non-LTR retrotransposons (2–3%). Uncharacterized repeats constitute a large proportion (16–20%) of the genome. The highly abundant repeat classes are satellites/low-complexity repeats, as well as short interspersed nuclear elements (SINEs) (5S-Deu-L21 class). However, at present, further studies are required to elucidate the role of repetitive sequences in regulation of gene expression and in evolution.

5.8 TAXONOMICALLY RESTRICTED GENES

Decoding the genome of any animal species, followed by comprehensive analysis of the component genes, invariably demonstrates that a considerable number of them (5–15%) encode proteins with no similarity to anything in the National Center for Biotechnology Information (NCBI) nr database. For example, *S. kowalevskii* and *P. flava* contain 1905 and 1551 gene models, respectively, with no detectable homologs (Oleg Shimakov, personal communication). Most of these genes are substantiated by corresponding mRNAs. Such genes are sometimes called "taxonomically restricted genes" (eg, Khalturin et al., 2009). In *Hydra*, it has been suggested that taxonomically restricted genes participate in the creation of phylum-specific novelties such as cnidocytes, in the generation of morphological diversity, and in the innate defense system. To understand deuterostome evolution and the origins of chordates, studies of the expression and function of these genes are essential, and additional experimental biology (especially physiological and pharmacological studies) and improved bioinformatics will also be required to fully understand and properly annotate them.

5.9 CONCLUSIONS

As of 2015, at least one genome has been decoded from each deuterostome phylum. These tell us that (1) genome sizes of echinoderms, hemichordates, and cephalochordates are comparable whereas those of urochordates are reduced, presumably because of their short generation times, although verte-brate genomes have been doubled; (2) a comparable level of synteny exists in genomes of hemichordates, cephalochordates, and vertebrates, and perhaps echinoderms as well; (3) although deuterostome genomes share gene families and exon-intron constitution, genomic comparisons reveal features common to all or specific to certain taxa. Genomic information provides a basis for further discussion of the origins and evolution of chordates. We have not yet fully char-acterized deuterostome genomes in the context of deuterostome evolution and chordate origins. Further investigations are needed to find out genomic changes associated with the chordate origins.

Chapter 6

The Origins of Chordates

To understand the origins of chordates, I have discussed evidence from the fossil record in Chapter 3, molecular phylogenomics in Chapter 4, and comparative genomics in Chapter 5. Do the hypotheses offered to explain the evolution of deuterostomes and the origins of chordates (Chapter 2) receive support from these disciplines? Further discussion of related issues may shed light on deuterostome evolution.

In Chapter 1, I discussed deuterostome taxa that should be considered to understand the origins and evolution of chordates. In Chapter 2, several hypotheses to explain chordate origins were introduced. Recent advances in deuterostome paleontology, molecular phylogenomics of metazoans and deuterostomes, and comparative genomics of deuterostomes were discussed in Chapters 3–5, respectively. The following chapters discuss recent advances in evolutionary developmental biology (Evo-Devo). Evo-Devo studies may provide information essential to further discussion of chordate origins. In Chapter 6, I first discuss the hypotheses and evolutionary scenarios mentioned in Chapter 2 and the support they receive from recent fossil data, molecular phylogeny, and genomics. Then, I discuss the pharyngeal and Hox gene clusters of deuterostomes as a genetic background to characterize them.

6.1 EVALUATION OF HYPOTHESES FOR CHORDATE ORIGINS

6.1.1 How Do We Interpret Deuterostome Evolution?

Before further discussion, I wish to explain my way of interpreting the evolutionary pathways of deuterostomes. Fig. 6.1 shows two general trees to explain phylogenetic relationships among deuterostomes with an evolutionary pathway leading to vertebrates. According to the first (Fig. 6.1A), a deuterostome ancestor gave rise to the ambulacraria, which diverged first from the main stem, forming its own lineage. Thereafter, cephalochordates branched off to form their own lineage, then urochordates, and finally vertebrates. In the second scheme (Fig. 6.1B), the deuterostome ancestral lineage first diverged into two main lines: one becoming the ambulacraria and the other the chordates. The former then divided into echinoderms and hemichordates. In the chordate lineage independent of the ambulacrarians, cephalochordates diverged first and then urochordates and vertebrates. I think the second is the most appropriate interpretation of the evolutionary history of deuterostomes. The reason why I do not support the Fig. 6.1A scenario is that it implies that chordates

Chordate Origins and Evolution. http://dx.doi.org/10.1016/B978-0-12-802996-1.00006-7

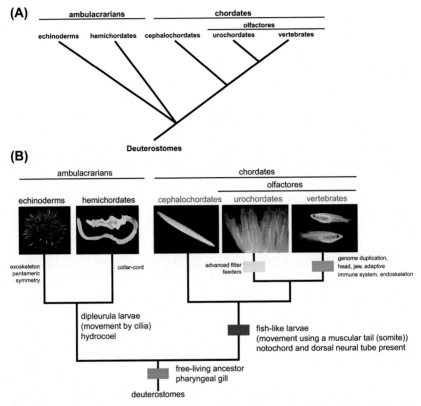

(A)

ambulacrarians — chordates — olfactores

echinoderms   hemichordates   cephalochordates   urochordates   vertebrates

Deuterostomes

(B)

ambulacrarians — chordates — olfactores

echinoderms   hemichordates   cephalochordates   urochordates   vertebrates

exoskeleton pentameric symmetry — collar-cord — advanced filter feeders — genome duplication, head, jaw, adaptive immune system, endoskeleton

dipleurula larvae (movement by cilia) hydrocoel — fish-like larvae (movement using a muscular tail (somite)) notochord and dorsal neural tube present

free-living ancestor pharyngeal gill

deuterostomes

FIGURE 6.1   **Two types of trees for deuterostome evolution.** (A) A tree tentatively called gradual evolution of ambulacrarians and chordates. (B) A tree tentatively called independent evolution of ambulacrarians and chordates. As discussed in the text, the latter provides better understanding of the evolutionary scenario.

emerged after the ambulacrarian divergence. An extrapolation of this idea leads to the interpretation that chordates originated downstream of ambulacrarians. Is this correct? Recent paleontological data as well as molecular phylogenomic data clearly indicate that ambulacrarians and chordates emerged almost simultaneously during the Cambrian Period (Figs. 3.2, 4.1, and 4.2). Or more precisely, paleontological data (Fig. 3.2) and molecular phylogenomic data (Fig. 4.2) tell us that the cephalochordate emergence was a little bit earlier than that of the Ambulacraria. That is, all five deuterostome taxa shared a common ancestor, from which ambulacrarians and chordates evolved contemporaneously, following their own evolutionary pathways. This idea is close to that proposed by Pat Willmer (1990) and sometimes cited as "the lawn theory" of metazoan evolution. I strongly support her concept. According to this line of thinking, chordates did not emerge downstream of ambulacrarians. Instead, we get hints about chordate evolutionary pathways from those of ambulacrarians, which evolved in parallel.

One very important question about the origin of deuterostomes is how their ancestor(s) appeared (ie, what they looked like); at present we have no idea at all. They may have been small animals with bilateral symmetry, creeping on the bottom of a shallow sea, judging from trace fossils of the very early Cambrian Period. Did they resemble extant acoels, or miniature, primitive acorn worms, or something else? The fossil record suggests that all deuterostome taxa emerged approximately 520 million years ago (MYA), and molecular phylogenomics data estimate their emergence times at 570 MYA (Fig. 4.2) or earlier (Fig. 3.2). Thus there is a gap of at least 50 million years (MY) between them. What happened during this geological time? If the generation time of a given deuterostome is hypothesized to be 3 years, 50 MY represents 16.6 million generations. This many generations might have been sufficient to allow the deuterostome ancestor to give rise to the five extant taxa. However, this answer raises the question as to how they all arose essentially simultaneously?

6.1.2 The Chordate Ancestor Was Free-Living

It is highly likely that deuterostome and chordate ancestors were free-living, similar to enteropneusts, rather than sessile, similar to pterobranchs. This is similar to the question of whether sessile ascidians or free-living cephalochordates emerged first. Molecular phylogenomics demonstrates that free-living cephalochordates diverged first among chordates (Figs. 4.1 and 4.2). Therefore it appears that our ancestor was free-living, rather than sessile, similar to extant ascidians.

6.1.3 The Auricularia Hypothesis Is Not Supported

The auricularia hypothesis, proposed by Garstang (1928a,b), was an attempt to explain how the chordate body plan originated from a deuterostome common ancestor by emphasizing the significance of changes in larval forms (Section 2.2; Fig. 2.3). A hypothetical dipleurula (auricularia-like) larva probably had an aboral band of feeding cilia near the mouth and a band of circumoral cilia across its lateral surface (Fig. 2.3A). Pterobranch-like, sessile animals with auricularia-like larvae led to the primitive ascidians through morphological changes in both larvae and adults. That is, pterobranch-like adults changed their feeding apparatus from external tentacles to internal branchial sacs. In larvae, the circumoral ciliated bands and their associated underlying nerve tracts moved dorsally to meet and fuse at the dorsal midline, forming a dorsal nerve cord in the chordate body (Fig. 2.3A). At the same time, the aboral ciliated band gave rise to the endostyle and ciliated tracts within the pharynx of the chordate. In Section 2.2, I explained that the hypothesis is still supported in revised form (eg, Nielsen, 1999). To explain further, Garstang (1928a,b) proposed two rounds of pedomorphosis during this evolutionary process: once in auricularia-like larvae of pterobranchs and once in tadpole-like larvae of ascidians (Fig. 2.5A).

First, since the chordate ancestor was likely a free-moving animal, pedomorphism is not always required to explain the emergence of a lancelet-like, free-living ancestor. However, the concept of pedomorphism will be discussed from a different point of view in relation to the mode of cephalochordate embryogenesis (Chapter 7). Second, as will be discussed in Chapter 7, when compared with deuterostome embryogenesis, neural tube formation of cephalochordate embryos is completely unrelated to the ciliary bands of ambulacrarian larvae, although a similar set of genetic tools is used for development of these different nervous systems (eg, Miyamoto et al., 2010). Neither the mode nor the timing of organ formation corresponds. The chordate neural tube is formed by curling up the edge of the dorsal ectoderm of the embryo, followed by fusion at the midline to form the tube. In addition, when the lancelet neurula-stage embryo is compared to an ambulacrarian embryo at the corresponding stage, it is evident that the ambulacrarian embryo never develops the ciliary bands, which are formed at later stages (Fig. 2.7). Therefore the auricularia hypothesis lost its standing in the face of growing evidence from deuterostome embryology. However, this challenge illustrates the value of comparative of deuterostome embryology.

6.1.4 The Calcichordate Hypothesis Cannot Be Accepted

Arguments against the calcichordate theory have already been discussed in Section 2.3; therefore they are mentioned only in passing here. This theory, proposed by paleontologist, Richard Jefferies, was rather recent (1996). Jefferies insists that the fossil echinoderm, *Mitrata,* which belongs to the Calcichordata, was the stem for the Olfactores (jointly called the Dexiothetes). The Olfactores subsequently lost biomineralization ability. Dexiothetism is a synapomorphy for the clade. This hypothesis has provoked heated controversies for the last decade, with diminishing support (Gee, 1996; Erwin et al., 2011).

First, molecular and genomic data indicate that echinoderms are a diverse group among deuterostomes by virtue of their biomineralization capacity. In hemichordate, cephalochordate, and urochordate genomes, there are no traces indicating the loss of genes involved in biomineralization. Second, gill slits found in the Calcichordata are not always specific to or shared by echinoderms and Olfactores. Instead, this character is a synapomorphy of deuterostomes (Section 6.2). Third, along with the concept discussed in Section 6.1.1, the comparison of adult ambulacrarian and chordate organs does not always provide clues to infer the origins of chordates. Jefferies's explanation of the transition from echinoderms to chordates includes various ideas, most of which are hard to accept. However, the aforementioned Calcichordata hypothesis highlights the significance of pharyngeal gills as a deuterostome-specific character and indicates the close phylogenic relationship between urochordates and vertebrates.

6.1.5 The Annelid Hypothesis Is Not Related to the Origins of Chordates

I also cannot support the annelid hypothesis in the context of chordate origins. As is evident from molecular phylogeny, protostomes (annelids and arthropods) are a discrete animal group, distinct from deuterostomes (ambulacrarians and chordates). Hints for resolving the problem of how chordates originated and evolved should be concealed among deuterostomes themselves. Because annelids and chordates are bilaterians, comparisons of annelid and chordate body plans might identify several morphological and anatomical similarities. Alternatively, similar to Hox cluster genes, arthropods and vertebrates share well-conserved expression and functional profiles of transcription factor genes. However, these similarities do not give us meaningful insights into the origins of chordates. This hypothesis may be useful in discussions of bilaterian evolution. I cannot support the axochord theory of the notochord origin either, which will be discussed in Chapter 9.

6.1.6 The Enteropneust Hypothesis and Inversion Hypothesis Need Reconsideration

The enteropneust hypothesis proposed by Bateson (1886) and others emphasized similarity (or homology) of adult organs and/or structures between enteropneusts and chordates. The proposed shared characters include the stomochord (and pygo-chord), corresponding to the chordate notochord, the collar cord (which he considered dorsal) corresponding to the chordate central nervous system (CNS), and pharyngeal gill slits in both groups. The pharyngeal gill will be discussed in Section 6.2 as a deuterostome-specific character. The relationship of the stomochord and collar cord will be discussed in Chapter 9. However, because ambulacrarians and chordates diverged early in the history of deuterostomes and because each of them selected an independent evolutionary pathway, comparison of adult organs does not always give us clues to infer the origins of chordates. Adult organ similarity should not be discussed relative to chordate origins but rather in relation to evolutionary pathways deuterostomes appear to have adopted.

The inversion hypothesis has been supported by molecular data—specifically, shared expression profiles of genes for signal pathway molecules and transcription factors with developmentally significant functions. The molecular inversion hypothesis originally showed that genes involved in dorsoventral axis formation are the same among arthropods and vertebrates (Arendt and Nübler-Jung, 1994; Holley et al., 1995). This hypothesis has been supported by many recent studies in this field. I cannot support the application of the molecular inversion hypothesis to the annelid hypothesis, simply because inversion would not promote formation of chordate-specific organs, such as the notochord (Chapter 8). I wish to apply the same strategy to the application of the molecular inversion hypothesis to the enteropneust hypothesis because this hypothesis also does not explain development of the chordate-specific characters. However, because there are similarities in adult organs between enteropneusts and chordates, at the moment I am hesitant to completely reject the new (inverted) enteropneust hypothesis, but evidence supporting its rejection will be discussed in Chapter 9.

6.2 THE PHARYNGEAL GENE CLUSTER AND THE ORIGIN OF DEUTEROSTOMES

To understand the origins of chordates, it is desirable to understand deuterostome origins themselves. As discussed in Section 1.2.1, deuterostomy includes (1) development of the blastopore into the anus and development of the mouth from a secondary opening, (2) radial cleavage, (3) indeterminate cleavage, (4) enterocoely, and (5) the tripartite composition of adult bodies (Table 1.1). In addition, pharyngeal gills (or gill slits) have sometimes been discussed as a synapomorphy of deuterostomes (Gillis et al., 2012; Graham and Richardson, 2012; Edlund et al., 2014). These structures are missing in extant echinoderms and amniotes. However, as the calcichordate hypothesis emphasizes, pharyngeal gills were present in stem echinoderms. The disappearance in amniotes is special evolutionary change that occurred only in this lineage. Recent decoding of acorn worm (hemichordate) genomes has revealed the genetic background of pharyngeal gills as a unique gene cluster that might characterize deuterostome genomes (Simakov et al., 2015).

Several independent studies have suggested syntenic organization of *Nkx2-1/4*, *Nkx2-2/9*, *Pax1/9*, *Slc25A21*, and *FoxA1/2* in the mouse genome (eg, Santagati et al., 2001) and the expression of *Pax1/9* in the pharyngeal gill region of acorn worms (Ogasawara et al., 1999). As discussed in Chapter 5, comparative genomics is a powerful tool to elucidate possible molecular mechanisms involved in evolutionary development. Analyses of acorn worm genomes reveal a microsyntenic cluster that is conserved in all deuterostome taxa except, as expected, for urochordate genomes. This is a block of six genes consisting of four transcription factor genes (*Nkx2.1*, *Nkx2.2*, *Pax1/9*, and *FoxA*) plus two nontranscription factor genes [*Slc25A21* (solute transporter) and *Mipoll* (mirror-image polydactyly one protein); Fig. 6.2A]. Nkx2.1 and Nkx2.2 are

FIGURE 6.2 **Conservation of a deuterostome pharyngeal gene cluster.** (A) Linkage and order of six genes, including four transcription factors (*Nkx2.1*, *Nkx2.2*, *Pax1/9*, and *FoxA*) and two nontranscription factors [*slc25A21* (solute transporter) and *mipoll* (mirror-image polydactyly one protein), which are putative bystander genes containing regulatory elements of *Pax1/9* and *FoxA*, respectively]. Pairing of *slc25A21* with *Pax1/9* and of *mipoll* with *FoxA* also occurs in protostomes, indicating bilaterian ancestry. This cluster is not present in protostomes, such as *Lottia* (lophotrochozoa), *D. melanogaster*, *C. elegans* (ecdysozoa), or in the cnidarian *Nematostella*. *Slc25A6* (the *slc25A21* paralog on human chromosome 20) is a potential pseudogene. Dots marking A2 and A4 indicate two conserved noncoding sequences first recognized in vertebrates and amphioxus and also present in *S. kowalevskii* and *A. planci*. (B) Expression of the pharyngeal gene cluster. Four transcription factor genes in a cluster are expressed in the pharyngeal/foregut endoderm of juvenile *Saccoglossus*. (a) *Nkx2.1* is expressed in a band of endoderm at the level of the developing gill pore, especially ventral and posterior to it (*arrow*), and in a separate ectodermal domain in the proboscis. It is also known as *thyroid transcription factor-1* because of its expression in the pharyngeal thyroid rudiment in vertebrates. (b and c) *Nkx2.2* is expressed in pharyngeal endoderm just ventral to the forming gill pore, shown in side view (*arrow* indicates gill pore) and ventral view. (d) *Pax1/9* is expressed in the gill pore rudiment itself. In the enteropneust, *S. kowalevskii*, this is its only expression domain, whereas in vertebrates it is also expressed in axial mesoderm. (e) *FoxA* is expressed widely in endoderm, but is repressed at the site of gill pore formation (*arrow*). (f) An external view of gill pores is shown; up to 100 bilateral pairs are present in adults, indicative of the large size of the pharynx. (*From Simakov, O., Kawashima, T., Marlétaz, F., Jenkins, J., Koyanagi, R., Mitros, T., Hisata, K., Bredeson, J., Shoguchi, E., Gyoja, F., Yue, J.X., Chen, Y.C., Freeman, R.M., Sasaki, A., Hikosaka-Katayama, T., Sato, A., Fujie, M., Baughman, K.W., Levine, J., Gonzalez, P., Cameron, C., Fritzenwanker, J.H., Pani, A.M., Goto, H., Kanda, M., Arakaki, N., Yamasaki, S., Qu, J., Cree, A., Ding, Y., Dinh, H.H., Dugan, S., Holder, M., Jhangiani, S.N., Kovar, C.L., Lee, S.L., Lewis, L.R., Morton, D., Nazareth, L.V., Okwuonu, G., Santibanez, J., Chen, R., Richards, S., Muzny, D.M., Gillis, A., Peshkin, L., Wu, M., Humphreys, T., Su, Y.H., Putnam, N.H., Schmutz, J., Fujiyama, A., Yu, J.K., Tagawa, K., Worley, K.C., Gibbs, R.A., Kirschner, M.W., Lowe, C.J., Satoh, N., Rokhsar, D.S., Gerhart, J., 2015. Hemichordate genomes and deuterostome origins. Nature 527, 459–465.*)

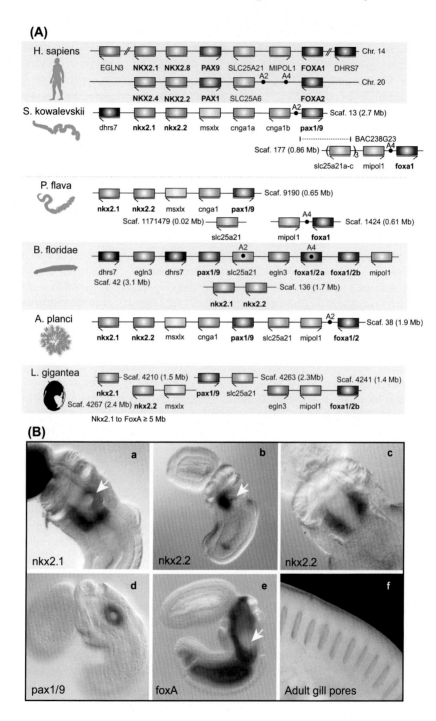

(A)

(B)

a nkx2.1

b nkx2.2

c nkx2.2

d pax1/9

e foxA

f Adult gill pores

NKX-homeodomain transcription factors that play critical regulatory roles in metazoan organ development (eg, Takacs et al., 2002). Pax1/9 is a paired domain-containing transcription factor that is involved in embryonic development. FoxA is forkhead domain transcription factor that plays a critical role in endoderm development. *Slc25A21* and *Mipol1* are bystander genes that contain, within their introns, regulatory elements for *Pax1/9* and *FoxA*, respectively. The cluster, from *Nkx2.1* to *FoxA1*, accounts for approximately 1.1 megabase-pairs (Mbp) of human chromosome 14 and 0.5 Mbp of the genome of the hemichordate *Saccoglossus kowalevskii* (Fig. 6.2A). In addition, the full ordered gene cluster also exists on a single scaffold in the genome of the crown-of-thorns starfish, *Acanthaster planci* (Fig. 6.2A).

Furthermore, the two noncoding elements, Slc25A21 and Mipol1, which are conserved among vertebrates and amphioxus, are found in hemichordate and *A. planci* clusters at similar locations (shown as A2 and A4 in Fig. 6.2A). The pairings of *Slc25A21* with *Pax1/9* and of *Mipol1* with *FoxA* occur also in protostomes, indicating bilaterian ancestry. However, the cluster is not present in protostomes such as *Lottia* (lophotrochozoa) (Fig. 6.2A), *Drosophila melanogaster* (ecdysozoa), *Caenorhabditis elegans* (ecdysozoa), or in the cnidarian *Nematostella*.

The *Pax1/9* gene, at the center of the cluster, is expressed in the pharyngeal endodermal primordium of the gill slit in hemichordates (Fig. 6.2B(d)), tunicates, amphioxus, fish, and amphibians; in the branchial pouch endoderm of amniotes (which do not complete the last steps of gill slit formation); and in other locations in vertebrates. *Nkx2.1* (thyroid transcription factor-1) is also expressed in the hemichordate pharynx endoderm in a band passing through the gill slit (Fig. 6.2B(a)), but not localized to a thyroid-like organ. The expression of *Nkx2.2* and *FoxA* in *S. kowalevskii* is also distinct, in that *Nkx2.2*, which is expressed in the ventral hindbrain in vertebrates, is expressed in pharyngeal ventral endoderm in *S. kowalevskii*, close to the gill slit (Fig. 6.2B(b and c)). *FoxA* is expressed throughout endoderm, but it is repressed in the gill slit region (Fig. 6.2B(e)). Coexpression of this cluster of the four ordered transcription factors during pharyngeal development strongly supports the functional importance of their genomic clustering. This cluster is now called "the pharyngeal gene cluster."

The presence of this cluster in the crown-of-thorns starfish, which lacks gill pores, and in terrestrial vertebrates, which lack gill slits, suggests that the cluster's ancestral role was in pharyngeal apparatus patterning as a whole, of which overt gill slits (perforations of apposed endoderm and ectoderm) were but one aspect. It also suggests that the cluster is retained in these cases because of its continuing contribution to pharyngeal development. The pharyngeal cluster has been implicated in long-range promoter–enhancer interactions, supporting the regulatory importance of this gene linkage. Alternatively, genome rearrangement in these lineages may be too slow to disrupt the cluster even without functional constraints. It may be concluded that clustering of the four ordered transcription factors and their bystander genes on the deuterostome stem served a regulatory role in evolution of the pharyngeal apparatus, the foremost morphological innovation of deuterostomes.

The pharyngeal gill is profoundly associated with filter feeding in deuterostomes (Fig. 6.2B(f)). Mucus produced by cells of this organ also plays an important role in nutrient capture in the filter-feeding system. Mucus-related genes are also likely to have evolved in relation to this system (Chapter 11).

6.3 HOX AND CHORDATE EVOLUTION

Hox genes, with their similar roles in animals as evolutionarily distant as humans and flies, have fascinated biologists since their discovery nearly 30 years ago (eg, Carroll et al., 2005). Hox genes comprise a subfamily of homeobox-containing transcription factors. In most metazoans studied thus far, Hox genes are clustered in the same genomic region, known as the Hox cluster. The Hox cluster shows spatial and temporal colinearity. That is, the expression patterns of Hox genes reflect their position in the cluster. Genes at the 3′ end are expressed in and pattern the anterior end of the embryo, whereas genes at the 5′ end pattern more posterior body parts. This phenomenon is called spatial colinearity. Moreover, gene position in the cluster also determines the onset of expression, with 3′ genes expressed at earlier developmental stages than those at the 5′ end. This phenomenon is called temporal colinearity. As a result of spatial and temporal colinearity, Hox genes are eventually expressed in a nested manner along the anterior-posterior axis of the animal body, resulting in a Hox code that bestows differential structural identity.

The Hox cluster of metazoans and its evolution have been reviewed by many researchers, including Garcia-Fernàndez (2005a,b), Duboule (2007), Holland (2013b), Pascual-Anaya et al. (2013), and Soshnikova (2014) (related references are therein). Among deuterostomes, cephalochordates are thought to have the most typical Hox cluster from the anterior-most *Hox1* to the posterior-most *Hox13*, although lancelets have also *Hox14* and *Hox15* (Amemiya et al., 2008; Holland et al., 2008; Fig. 6.3A). Recent studies reveal the presence of a well-organized Hox cluster in the Ambulacraria (Freeman et al., 2012; Baughman et al., 2014), whereas the urochordate Hox cluster, especially that of the larvacean, *Oikopleura dioica*, is disorganized (Shoguchi et al., 2008; Denoeud et al., 2010). Vertebrates possess multiple clusters as a result of two rounds of genome-wide gene duplication (2R-GWGD), resulting in the four paralogous Hox clusters of jawed vertebrates (HoxA, B, C, and D clusters; Fig. 6.3A; eg, Hoegg and Meyer, 2005). In addition, teleost fishes experienced an additional third round, resulting in up to seven or eight Hox clusters (eg, Crow et al., 2006).

In vertebrates, changes in the Hox code may promote evolutionary novelties, such as the fin-to-limb transition (eg, Ahn and Ho, 2008), the number of vertebrae (eg, Burke et al., 1995), the snake body plan (Di-Poï et al., 2010), or the presence or absence of ribs in the trunk (Wellik and Capecchi, 2003), and so forth. Although recent studies have accumulated data to answer the question of

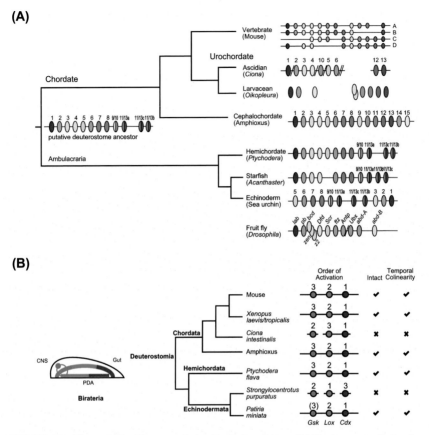

FIGURE 6.3 Hox and ParaHox gene clusters in deuterostomes. (A) Hox gene organization in deuterostomes. That of *Drosophila* is shown as a reference. *Colored ovals* indicate Hox genes. Genes of smaller paralogous subgroup numbers are to the left and those of larger numbers are to the right. A putative chordate ancestor may have possessed a single Hox gene cluster with approximately 13 genes. This would have been conserved in hemichordates and cephalochordates, although cephalochordates must have duplicated the posterior-most genes. Vertebrates have four Hox gene clusters (A–D), from the top to the bottom. Echinoderms most likely lost *Hox6*. In urochordates, the Hox cluster was reorganized; in *C. intestinalis*, Hox genes are mapped on two chromosomes and the putative gene order is shown for *Hox2-4* and *Hox5-6*. With respect to *Oikopleura* Hox genes, the chromosomal location is unknown. *(Modified from Freeman, R., Ikuta, T., Wu, M., Koyanagi, R., Kawashima, T., Tagawa, K., Humphreys, T., Fang, G.C., Fujiyama, A., Saiga, H., Lowe, C., Worley, K., Jenkins, J., Schmutz, J., Kirschner, M., Rokhsar, D., Satoh, N., Gerhart, J., 2012. Identical genomic organization of two hemichordate hox clusters. Current Biology 22, 2053–2058.)* (B) The ParaHox cluster of deuterostomes. Conserved synteny of three ParaHox genes (*Gsx*, *Lox*, and *Cdx*) and two other genes is shown. *(From Garstang, M., Ferrier, D.E., 2013. Time is of the essence for ParaHox homeobox gene clustering. BMC Biology 11, 72.)*

how the Hox code promoted morphological changes, especially in the development of the CNS during deuterostome evolution, it still remains to be elucidated how the putative Hox repertoire of the last common ancestor was modified in different deuterostome taxa to generate morphological novelty.

6.3.1 Ambulacrarian Hox Genes

As discussed in previous chapters, among ambulacrarians, hemichordates are thought to retain more features of the last common ancestor than echinoderms. A recent study identified the presence of a single Hox cluster in genomes of two enteropneusts, *S. kowalevskii* and *Ptychodera flava* (Fig. 6.3A; Freeman et al., 2012). Their Hox clusters show identical organization, with 12 Hox genes arrayed in approximately 500 kb, all with the same transcriptional orientation except for the terminal pair of Ambulacrarian-specific posterior Hox genes, *AmbPb* and *AmbPc* (previously named *Hox11/13b* and *Hox11/13c*, respectively; Freeman et al., 2012; Fig.6.3A). Overall, the hemichordate Hox cluster reflects a prototypical organization among deuterostomes.

The expression profile of Hox cluster genes in juvenile *S. kowalevskii* suggests the role of Hox cluster genes during deuterostome evolution (Chapter 9). From *Hox1* to *Hox11/13a, b, c* in this order, the 12 genes are expressed along the anteroposterior axis in the epidermis of the juvenile trunk (Fig. 9.1). Expression domains are broadly conserved between hemichordate epidermis and chordate CNS (Fig. 9.1).

The features of echinoderm Hox clusters have also been recently described (Sea Urchin Genome Sequence Consortium et al., 2006; Cameron et al., 2006). First, the Hox cluster of the sea urchin, *Strongylocentrotus purpuratus*, is a single cluster of approximately 600 kb that contains 11 Hox genes (*Hox4* is missing; Fig. 6.3A). It also appears reordered because *Hox1-3* are near the posterior end of the cluster (Fig. 6.3A). In addition, sea urchin Hox genes are not expressed during embryogenesis but are expressed during juvenile development. These data suggest to some sea urchin embryologists that this Hox shuffling may be associated with echinoderm-specific, pentameric symmetry, although their function remains to be elucidated. However, it has been shown that the starfish, *A. planci*, has an organized Hox cluster of 11 genes, in which *Hox6* is missing (Fig. 6.3A; Baughman et al., 2014). Because a starfish with pentameric symmetry retains an organized Hox cluster, the relationship between Hox rearrangement and the pentameric symmetry requires further investigation.

Crinoids (*Metacrinus rotundus*) and asteroids (*Asterina minor* and *Patiriella exigua*) have a *Hox4* gene (eg, Hara et al., 2006). It remains to be seen whether the loss of *Hox4* in sea urchins occurred after the split of echinoids, holothuroids, and ophiuroids from other echinoderms or whether sea urchins actually lost *Hox6* rather than *Hox4* because of confusion of these two Hox genes.

6.3.2 Cephalochordate Hox Genes

Cephalochordates occupy a key phylogenetic position close to the ancestral condition of chordates, and they provide a valuable outgroup for evolutionary and comparative studies of vertebrates (Bertrand and Escriva, 2011). *Branchiostoma floridae* and *Branchiostoma lanceolatum* possess the most prototypical Hox cluster identified so far in deuterostomes (Fig. 6.3A). It contains 15 Hox genes, including *Hox14* and *Hox15*, the largest Hox clusters hitherto reported, spanning approximately 470 kb, and all in the same transcriptional orientation. For these reasons, the amphioxus Hox cluster appears to reflect the ancestral chordate condition, with counterparts for every paralogous group (PG) of vertebrates.

The posterior genes are the subject of some debate. Because phylogenetic trees do not show clear orthologous relationships between posterior genes from different deuterostome phyla, this phenomenon was thought to be a consequence of the higher evolutionary rate of the posterior part of the cluster, the so-called deuterostome posterior flexibility (Ferrier et al., 2000). However, an alternative scenario is that some posterior genes originated independently in different lineages by tandem duplication events. In relation to the uncertainty about the posterior part of the Hox cluster, disruption of colinearity for some amphioxus Hox genes was reported, the most striking example of which was that *Hox14* was expressed in cerebral vesicles (Pascual-Anaya et al., 2012). Interestingly, forebrains of other animals are characterized as Hox-negative regions, making amphioxus a surprising oddity and raising the possibility that the amphioxus Hox cluster and its regulation are not as prototypical as previously thought.

6.3.3 Urochordate Hox Genes

As previously discussed, urochordate genomes are highly divergent, and *Ciona intestinalis* possesses a rather atypically organized set of Hox genes (Spagnuolo et al., 2003; Ikuta et al., 2004). The *C. intestinalis* Hox cluster is divided into two groups, located on different chromosomes (Shoguchi et al., 2008). These groupings are *Hox1-6* and *10* on chromosome 1 and *Hox12-13* on chromosome 7 (Fig. 6.3A). In addition, *Hox7-9* and *11* are absent in all ascidians sequenced thus far (Dehal et al., 2002). Furthermore, two independent translocation events disrupted the order of Hox genes on chromosome 1. *Hox10* is located between *Hox4* and *Hox5*, and the Hox-related gene, *EvxA*, sits between *Hox1* and *Hox2*. In fact, some authors consider that sensu stricto, *Hox1-10* are not linked because they span approximately 5 Mb (Ikuta et al., 2004). Another urochordate, the larvacean, *O. dioica*, shows a further dramatic disintegration of the cluster with Hox genes scattered throughout the genome (Seo et al., 2004). Only two genes, *Hox9a* and *Hox9b*, are linked, probably as a result of a species-specific tandem duplication event. Moreover, larvaceans seems to have lost *Hox3, 5, 6, 7,* and *8*.

Surprisingly, even with such a gene shuffling and deletion, most ascidian Hox genes are expressed in somewhat colinear fashion in the CNS (Ikuta et al., 2004; Fig. 7.7B). In addition, as in *C. intestinalis*, *O. dioica* Hox genes were expressed with partial spatial colinearity in the CNS.

6.3.4 Vertebrate Hox Genes

Because of 2R-GWGD, Hox clusters in vertebrates increased to four PGs: HoxA to HoxD (Fig. 6.3A). In addition, teleost fishes expanded to seven or eight clusters, arising from a teleost-specific third round. Salmonids have up to 13 clusters after an additional salmonid-specific fourth round (Mungpakdee et al., 2008). In all such GWGD events, duplication of the Hox cluster was followed by various Hox gene losses, eventually resulting in unique combinations of Hox genes in different groups, similar to a bar code (a "genomic Hox-bar code"; eg, Duboule, 2007). Accordingly, it should be possible to determine to which group a genome of unknown origin belongs just by analyzing the Hox gene/cluster content.

6.3.4.1 Cyclostome Hox Genes

Cyclostome Hox clusters have been discussed frequently with regard to when the second round of GWGD occurred after the split of vertebrates from the chordate stem (Soshnikova, 2014). A conclusive picture of hagfish Hox genes and clusters is far from definitive because genomic organization of hagfish Hox genes remains a mystery (Stadler et al., 2004). A total of 25 Hox genes were found, but only two Hox clusters were recognizable: cluster *Pm1Hox* (with *Hox2, 3, 4, 5, 6, 7, 8, 9,* and *11*) and *Pm2Hox* (*Hox1, 4, 5, 7, 8, 9, 10,* and *11*). Surprisingly, 44 distinct lamprey Hox genes have been reported, although assignments of some lamprey central Hox genes to PG5–7 remain uncertain. Interestingly, no Hox gene belonging to PG12 has been found in any of these studies, suggesting the possible loss of all *Hox12* genes in the last common ancestor of lampreys. At present it is not clear whether lampreys possess representatives of the four gnathostome Hox clusters or whether some originated by independent duplication events (eg, if five clusters are confirmed). Therefore it may be that there is a gap in Hox genes between cyclostomes and gnathostomes.

6.3.4.2 Gnathostome Hox Genes

The two main groups of gnathostomes are the chondrichthyans (cartilaginous fishes) and the osteichthyans (bony fishes). Interestingly, elasmobranchs seem to have completely lost the HoxC cluster (King et al., 2011). On the other hand, the Hox clusters of osteichthyans are the most intensively studied among all deuterostomes.

If the Hox cluster of the last common ancestor consists of 12 or 13 genes, 2R-GWGD would imply 48 or 52 homeobox genes in vertebrates. Comparison of Hox inventories of different tetrapods allows us to infer a tetrapod ancestral condition of up to 41 Hox genes (Liang et al., 2011) and an amniote ancestral condition of 40 Hox genes (after the loss of *HoxC1*), the full set of which only the green anole (*Anolis carolinensis*) retains, whereas mammals and chickens have independently lost *HoxC3*. Although *Xenopus tropicalis* has 38 Hox genes, the amphibian ancestor probably had 40 genes after losing *HoxD12*. A surprising discovery was that the Atlantic salmon, *Salmo salar,* and the rainbow trout, *Oncorhynchus mykiss*, have 13 Hox clusters, arising from a salmonid-specific fourth round of GWGD (Mungpakdee et al., 2008). *Salmo* has 118 Hox genes plus 8 pseudogenes, the largest Hox repertoire to date.

6.3.5 Deuterostome Posterior Hox Genes: An Unsolved Origin

In contrast to conservation of the anterior and middle classes of Hox genes, the posterior Hox genes have changed differently in different deuterostome lineages. The ambulacrarian posterior *AmbP* genes, *Hox11/13a, b,* and *c,* form an independent clade (Freeman et al., 2012). This may result from a higher rate of evolution of posterior Hox genes, precluding their grouping in phylogenetic trees. In addition, the amphioxus *Hox9–12* and *Hox13–15* clades usually group with vertebrate *Hox9–10* and *Hox11–14,* respectively, suggesting an independent origin for these genes. This would imply the presence of at least one ancestral *PG13/14* gene in chordates, from which the amphioxus *Hox15* and vertebrate *PG13* and *PG14* genes originated. Therefore the last common ancestor of chordates had at least three posterior Hox genes: one *PG9/10;* one *PG11/12,* from which amphioxus *Hox13–14* originated; and one *PG13/14,* from which amphioxus *Hox15* originated. The two latter probably come from a *PG11/14* ancestral gene because all chordate *Hox11–15* genes form a monophyletic clade, implying a primordial condition of two ancestral genes that quickly expanded into a three-gene condition arising from tandem duplication.

Although the last common ancestor of deuterostomes likely had the PG9/10 ancestral gene, the origin of *AmbP* remains unsolved. Hox genes have generally only one intron, splitting the gene into two exons, the second one containing the homeobox. However, some posterior Hox genes possess a second intron, splitting the homeobox into two exons. These are the vertebrate *Hox14* genes; lamprey *Hox13β;* and amphioxus *Hox11, Hox12,* and *Hox14.* Ambulacrarian posterior Hox genes lost this intron in the last common ancestor whereas in vertebrates, the two or three ancestral posterior Hox genes contained this intron, which subsequently was lost independently in the amphioxus and vertebrate lineages. Finally amphioxus *Hox12* secondarily acquired a different second intron; therefore the posterior Hox genes are more flexible than central and anterior Hox genes.

6.4 PARAHOX GENES

The ParaHox gene cluster consists of *Gsx, Xlox,* and *Cdx* and has been proposed as the paralog or evolutionary sister of the Hox gene cluster (Brooke et al., 1998; Garcia-Fernàndez, 2005b; Garstang and Ferrier, 2013). The ParaHox cluster is well conserved in metazoans (Fig. 6.3B) and is involved in anterior-posterior development of the nervous systems and guts of animals. The cluster displays spatial and temporal collinearity, as in the case of the Hox cluster. Characterizations of ParaHox clusters in the sea star *Patiria maniata* (Annunziata et al., 2013), the hemichordate *P. flava* (Ikuta et al., 2013), and the cephalochordate *B. floridae* (Brooke et al., 1998) suggest tighter temporal than spatial colineality, with intact, ordered clusters. As in the case of Hox clusters, further studies are required to understand how the ParaHox cluster contributed to the evolution of deuterostomes and the origins of chordates.

6.5 CONCLUSIONS

In this chapter I first examined the validity or plausibility of hypotheses to explain the evolutionary scenario leading to chordates. Given the earliest divergence of cephalochordates and with a free-living chordate ancestor, I provided evidence that most of the hypotheses, including the auricularia hypothesis, the calcichordate theory, the annelid hypothesis, and the enteropneust hypothesis, are not supported by recent data from paleontology, molecular phylogeny, or comparative genomics. In the following chapters, the inverted enteropneust theory will be discussed in detail with the new organizers hypothesis of chordate origins.

Molecular evidence indicates that the pharyngeal gill is a synapomorphy of deuterostomes, a discrete character for separating them from protostomes. The Hox cluster is conserved among metazoans, and the deuterostome ancestor likely had a cluster of 12 homeobox genes. It is evident that the original cluster has experienced different changes in different taxa. An interesting finding is that hemichordate Hox genes are expressed along the anteroposterior axis in epidermal cells of juveniles, a similar pattern to that found in the vertebrate CNS, suggesting the proximity of hemichordates and chordates.

Chapter 7

The New Organizers Hypothesis for Chordate Origins

What is the most plausible explanation for mechanisms involved in the origins and evolution of chordates? Most chordate-specific characters are associated with the emergence of a fish-like larva that locomotes efficiently by beating its tail. Comparisons of embryogenesis in cephalochordates and hemichordates offer insights about chordate origins. The new organizers hypothesis emerges from this comparison.

In Chapter 6, I reexamined several hypotheses proposed for the origins of chordates in light of recent developments in molecular phylogenetics and comparative genomics of deuterostomes (Table 5.1). Most of the hypotheses are poorly supported, or even negated, by recent evidence, although some include concepts that are useful for further discussion (Chapters 8 and 9). One reason for this is that most hypotheses are based on comparisons of adult morphology and anatomy and fail to consider the importance of embryonic development. Another reason is that some hypotheses mistakenly compare larval morphology to that of adults or vice versa. Deuterostomes generally adopt a biphasic life history, larvae and adults, with metamorphosis linking the two phases. This biphasic lifestyle offers two opportunities for a given species to invade new niches. Usually adult forms exhibit major differences compared with larval forms, and larval forms are more conservative than adult forms, presumably because larvae must seek sites that will sustain them as adults, although some species develop two or more larval forms.

On the basis of criticisms of previously proposed hypotheses, this chapter first discusses what chordates are. Several features have been proposed to distinguish chordates, but I will examine the definitive features carefully once again. In doing so I will outline the chordate evolutionary story. I will propose the new organizers hypothesis of chordate origins, which is a refined version of the aboral-dorsalization hypothesis I previously proposed (Satoh, 2008).

7.1 CHORDATE FEATURES

As discussed in Chapter 1, chordates are easily distinguished from other deuterostomes by morphological features in larval and adult stages. Those widely

Chordate Origins and Evolution. http://dx.doi.org/10.1016/B978-0-12-802996-1.00007-9

agreed to characterize chordates are the notochord; the dorsal, hollow neural tube (it should be emphasized that the tube is structurally different from the cord, based on embryogenesis, and discussed later); somites or myotomes; pharyngeal gills (slits); the endostyle; and the postanal tail. Of these, pharyngeal gills are common to all deuterostomes (Section 6.2). Therefore these emerged before the divergence of chordates and ambulacrarians. The endostyle is a longitudinal, ciliated organ on the ventral wall of the pharynx that secretes mucus used by filter feeders to trap food (see Section 7.10). The endostyle is seen in cephalochordates, urochordates, and the ammocete larvae of lampreys. During lamprey metamorphosis, the endostyle is transformed into the adult thyroid gland.

The notochord is a rod-shaped organ found in all chordate embryos that provides the stiffness required for efficient caudal muscle function (Figs. 1.4 and 1.5; eg, Satoh et al., 2012). Chordates were named after this organ because it is the most prominent larval feature. The notochord also defines the craniocaudal axis of the fish-like larvae. For cephalochordates, the notochord persists throughout life and is essential for larval and adult locomotion. For larvacean urochordates, the notochord also persists throughout life, whereas in ascidians the notochord exists during the larval free-swimming phase but degenerates during metamorphosis and disappears in sessile adults. In lampreys (agnathans) and in primitive fish, such as sturgeons, the notochord persists throughout life (eg, Mansfield et al., 2015). However, in higher vertebrates the notochord becomes ossified in vertebrae and contributes to the nucleus pulposis in the centers of intervertebral discs.

Myotomes are anatomically defined as groups of muscles that are served by a spinal nerve root. Myotomes consist of blocks of striated muscle cells, very prominent along the lancelet trunk (Fig. 1.4). They are evident in the fossil, *Pikaia* (Fig. 3.4C). Somites are embryonic regions that contribute to development of muscles. In the context of chordate origins, somites and myotomes have similar import, namely to provide faster and more efficient larval locomotion. It is sometimes argued that somites were lost in urochordates. However, urochordates, such as ascidian larvae, develop well-organized muscle structure, which provides power for efficient tail-beating locomotion, comparable to that of other chordates (Fig. 1.5). The postanal tail is easily seen in adult cephalochordates and fishes because the anus is located in the ventral trunk slightly anterior to the base of the tail (Fig. 1.4).

The dorsal, hollow, neural tube is unique to chordates (eg, Holland, 2009). In chordate embryos, presumptive cells of the central nervous system (CNS) separate from ectodermal epidermal cells, flatten on the dorsal side to form the neural plate, and then the two edges move dorsally to create a closed neural tube. This mode of neural tube formation does not occur in nonchordate deuterostomes or protostomes. The significance of neural tube formation should be emphasized in relation to chordate evolution. The vertebrate brain marks the most complex neural network structure. The complexity of the vertebrate brain has been achieved by numerous folds of the neural epithelium, where cells proliferate and migrate to the interior to form several layers. The establishment of this complexity would not be possible if the basic configuration was not a tube. Its complexity is different from that of the insect central ganglion, or the collar cord of enteropneust hemichordates, the embryological or adult organogenic antecedents of which are not tubular. Neural tube formation is strictly a chordate evolutionary innovation.

7.2 THE NEW ORGANIZERS HYPOTHESIS OF CHORDATE ORIGINS

Of the six defining chordate characters, the pharyngeal gill is common to all deuterostomes, and in chordate taxa, both the pharyngeal gill and the endostyle are associated with filter feeding. Although the endostyle is a chordate novelty, it is not likely the main structural basis for chordate origins.

In contrast, four other features are apparently associated with the advent of fish-like larvae. Before discussing these novel features, we need to reconsider the use of the term *tadpole-like larvae*. There are two reasons why this term has been frequently used. First, ascidians are thought to be the earliest group to diverge from the chordate stem (Section 2.4) and their larvae are tadpole-like. Second, during the late 19th and early 20th centuries, when chordate evolution was debated heavily, the main deuterostome species used for embryology were frogs and sea urchins, or sometimes ascidians. Therefore tadpoles were embryological models for studying not only ontogeny but also phylogeny. However, it is now evident that cephalochordates diverged first and their larvae are fish-like. Therefore the term *fish-like larvae* sounds more descriptive of larval morphology for chordate origins. As is seen in cephalochordates and many vertebrates, fish-like larvae enable a developmental strategy that involves settling new habitats in which to live as adults. Because this adaptation was so effective, it has been retained for more than 500 million years (MY). Therefore the morphology of fish-like larvae does not change much during metamorphosis to achieve adult morphology. The reduced morphological change during metamorphosis has led to the misunderstanding that they are direct developers, but it not so. This term also has several implications for chordate evolution and the origin of vertebrates, as discussed below. I will use the term *fish-like larvae* hereafter.

Returning to chordate features, the notochord, somites (myotomes), postanal tail, and dorsal hollow neural tube are all novelties associated with the occurrence of fish-like larvae (Satoh, 2008; Satoh et al., 2012, 2014a). The Ambulacraria and all marine, nonchordate invertebrates develop larvae that swim with cilia or ciliary bands, as the pluteus larvae of sea urchins and the tornaria larvae of acorn worms (Figs. 1.2 and 1.3). In contrast, chordate fish-like larvae swim with their tails (Figs. 1.4 and 1.5). This movement facilitates faster locomotion than cilia-based movement and is advantageous for capture prey. The notochord provides the stiffness required for efficient caudal muscle function. The concept of "new organizers" explains these novel structures. That is, chordate-specific structures were developed, presumably rather independently, by virtue of organizers that were involved in the formation of each structure.

A distinct difference between the dorsoventral (DV)-axis inversion hypothesis and the new organizers hypothesis is that the former appears to fail to explain how the notochord, the somites, and the dorsal neural tube ever developed because there are no corresponding structures in early ambulacrarian embryonic stages (Section 2.7; Fig. 2.7). This is evident when chordate embryos are compared with ambulacrarian embryos at corresponding stages (Fig. 2.7). That is, these structures only appear in chordates. The original aboral-dorsalization hypothesis emphasized the location in which novel structures are formed (Section 2.7) and explained that novel structures had to be formed on the aboral side of the early ambulacrarian embryo or the dorsal side of the early chordate embryo because of spatial constraints upon the development of novel structures. This line of thought will be discussed further in Chapter 8 in relation to embryonic axis formation in deuterostomes.

7.3 CEPHALOCHORDATE EMBRYOGENESIS: PRIMITIVE CHORDATE BODY-PLAN FORMATION

Early embryogenesis of amphioxus, up to the late neurula stage, provides us with various insights into chordate origins (Hatschek, 1893; Conklin, 1932; Hirakow and Kajita, 1994; Whittaker, 1997; Holland and Holland, 2007; Nielsen, 2012; Jandzik et al., 2015) (Figs. 1.4C and 7.1). Here I wish to introduce its key features. As discussed earlier, cephalochordates represented the earliest divergence from the chordate stem. Genomic investigations indicate that lancelets have the basic chordate genetic repertoire, which likely leads directly to vertebrates. The Cephalochordata comprises only 35 or so species in 2 genera, having similar morphology (Poss and Boschung, 1996). In addition, the morphology of lancelet-like fossils, including *Pikaia* (Conway–Morris, 1982) and *Cathaymyrus* (Shu et al., 1996), is similar to that of extant lancelets, indicating that further morphological innovation has not occurred for more than 500 MY. Lancelet embryology also appears to have been highly conserved.

In lancelets, radial cleavage forms a hollow blastula strongly resembling the cleavage and hollow blastula of sea cucumbers (echinoderms) and acorn worms (hemichordates; Fig. 7.1A; also compare Fig. 1.4C to Figs. 1.2C and 1.3C). Their chordate affinities appear during the early gastrula stage. In contrast to ambulacrarian gastrulation, in which the archenteron invaginates into a wide blastocoelic space (Figs. 1.2C and 1.3C), cephalochordate gastrulation occurs as the flattened endodermal plate ingresses or sinks deeply to the animal side of the embryo, eliminating the blastocoelic space and causing the embryo to become cup-shaped with a deepened archenteron (Fig. 7.1B and C and Fig. 1.4C). Urochordates show a similar pattern of gastrula formation (Fig. 1.5G and H). This is the first stage of embryogenesis in which differences between ambulacrarians and chordates become evident. In chordate embryos, presumptive dorsal-side cells are slightly thicker than presumptive ventral-side cells (Fig. 7.1C). No such difference is seen in ambulacrarian early gastrulae. This provides an important clue for understanding the presence of a genetic device or novel gene regulatory network (GRN) that produces a chordate-specific body plan, which will be discussed further in Chapter 8.

By the late gastrula stage, the embryo has become ovoid and slightly flattened, and the neural plate is formed along the flattened dorsal side of the embryo (Figs. 7.1D and 1.4C). Ectodermal cells have each developed a single cilium, as in ambulacrarian gastrulae at the same stage, and the gastrula begins to rotate within the perivitelline space. During the neurula stage, the embryo hatches and the free-living larva locomotes using cilia. At metamorphosis, the animals begin swimming using their tails. As in vertebrates, neurulation begins with enclosure of the neural plate. As both ectodermal edges move toward the dorsal midline and fuse, the neural plate becomes distorted below into a more pronounced V-shape (Holland and Holland, 2001). The V-shaped ridges of the sunken neural plate gradually coalesce from the rear forward, creating a closed neural tube (Fig. 7.1E and F). During neural tube formation, inside of the embryo, left and right dorsal sides of the archenteron invaginate or pouch off (Fig. 7.1E and F), followed by pouching off of midline roof cells (Fig. 7.1F and G). The lateral pouches give rise to the somites whereas the roof pouch creates the notochord (Fig. 7.1E–G).

Relative to chordate origins, cephalochordate embryogenesis, as described earlier, appears intermediate to some extent between the embryogenesis of the ambulacrarian and chordate clades. First, larval ciliary locomotion is an ambulacrarian trait, although it is replaced later by lateral undulations of the muscular tail. Second, the hollow neural tube is formed dorsally soon after gastrulation. Third, the cephalochordate notochord is formed by pouching-off of the dorsal region of the archenteron, as are the somites. Interestingly, the cephalochordate notochord displays muscle-like properties (Ruppert, 1997). Expressed sequence tag analysis indicated that approximately 11% of genes expressed in the *Branchiostoma* notochord encode muscle proteins, including actin, tropomyosin, troponin I, and creatine kinase (Suzuki and Satoh, 2000; Urano et al., 2003). It should be mentioned that in lampreys, the notochord is formed as a median fold from the dorsal roof of the archenteron, as in the case of amphioxus. On the other hand, in urochordates and gnathostomes, the notochord is formed by the convergent extension of cells that are bilaterally positioned in the gastrula (Fig. 7.2A; eg, Munro and Odell, 2002). They never possess muscular properties, and vacuolation of these cells increases cell volume and stiffness (eg, Jiang and Smith, 2007). Here, once again, it is worth mentioning that formation of the notochord and the neural tube occurs during the neurula stage, not the juvenile stage, if the stomochord/pygochord and collar cord are assumed to correspond to the notochord and neural tube, respectively. Again, it is highly likely that the notochord and the neural tube are novel and specific to chordates. One more interesting suggestion from lancelet embryogenesis is that formation of the neural tube clearly occurs

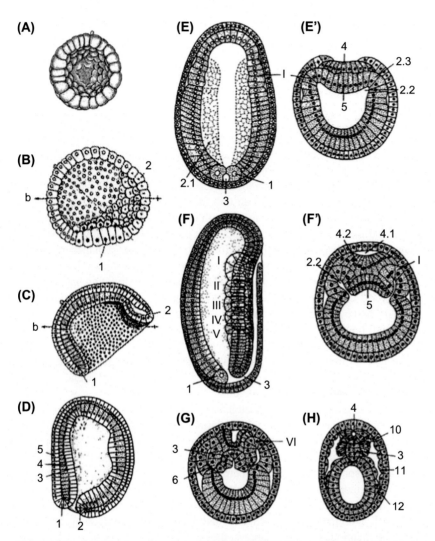

FIGURE 7.1 Amphioxus embryology. (A) Blastula; (B) initial gastrula; (C) mid-gastrula; (D) mid-to-late gastrula; (E, E′) initial neurula, sagittal section (E) and cross-section (E′); (F, F′) mid neurula, longitudinal section (F) and cross-section (F′); (G) mid-to-late neurula; and (H) late neurula. The radially symmetrical cleavage pattern of amphioxus to form a hollow blastula resembles that of acorn worms (hemichordates) (A), and their chordate affinities appear during the gastrula and neurula stages (B–H). In cephalochordates, gastrulation occurs as the flattened endodermal plate ingresses to the animal pole of the embryos, and this morphogenetic movement eliminates the blastocoelic space so that the embryo becomes cup-shaped with a deepened archenteron (B–D). By late gastrula stage, the embryo has become ovoid and slightly flattened, and the neural plate is formed as the flattened dorsal side of the embryo (D). (B–D) b indicates the anterior–posterior axis; 1, endoderm; 2, mesoderm; 3, mesodermal groove; 4, notochord; 5, neural plate. As in vertebrate embryos, neurulation begins with enclosure of the neural plate (E–H). During neural tube formation, the notochord develops from the adjacent chordamesodermal plate that constitutes the

before that of the notochord (Fig. 7.1). This suggests that the notochord itself did not induce neurulation in the original phase of chordate embryogenesis, even if the dorsal midline archenteron did have this potential.

7.4 CHORDATE FEATURES AND MOLECULAR DEVELOPMENTAL MECHANISMS

As discussed previously, I believe that the occurrence of fish-like larvae with caudal locomotion is a key evolutionary event in chordate evolution. As the most prominent feature of chordates, the notochord and dorsal CNS, their development, and the genes involved in their formation have been most intensively studied and discussed. However, even if the notochord and dorsal CNS were formed in a given metazoan, it cannot unequivocally be said that it was a chordate because chordates developed many novel characters. In other words, the chordate body plan is an accomplishment involving combinatorial innovation of the three germ layers, ectoderm, mesoderm, and endoderm. Therefore we should pay much more attention to genetic changes involved in modification of organs/tissues originating from each germ layer. Hereafter I describe the molecular developmental biology of chordate-specific characters mentioned previously, essential knowledge for further discussion of chordate origins.

7.5 THE NOTOCHORD: A MESODERMAL NOVELTY

From an evolutionary point of view, the primary function of the notochord is to provide stiffness for efficient tail muscle function. In urochordates and gnathostomes, the notochord is formed by convergent extension of the precursor cells that are bilaterally positioned in the early embryo (Fig. 7.2A; Munro and Odell, 2002). Convergence, intercalation, and extension of notochord cells are among the most significant morphogenetic events along the dorsal midline of chordate embryos. At a later stage, the notochord is surrounded by a thickened basement membrane called the notochordal sheath. In vertebrates, the embryonic region that gives rise to the notochord serves as a signaling center, or "Spemann's organizer," and induces a secondary axis when transplanted into another embryo (Spemann and Mangold, 1924). Chordin, Noggin, and Follistatin function as the organizing signals (Heasman, 2006; Gilbert, 2013). In addition, the

roof of the archenteron (F′–H). That is, the notochord is formed by an upward pouching-off of midline cells along the chordamesodermal plate (F–H). (E, F) I–V indicate the number of somites; 1, posterior polar cells; 2.1, mesodermal band; 2.2, mesodermal groove; 2.3, mesodermal fold; 3, blastopore; 4, neural plate; 4.1, neural cavity; 4.2, neural plate; 5, notochord primordium. (G, H) 3, notochord; 4, neural tube; 6, mesodermal groove; 10, lateral plate; 11, body wall; 12, visceral wall. *From Hatschek, B., 1893. The Amphioxus and its Development. Swan Sonnenschein & Co., London; Conklin, E.G., 1932. The embryology of amphioxus. Journal of Morphology 54, 69–151; Hirakow, R., Kajita, N., 1994. Electron microscopic study of the development of amphioxus,* Branchiostoma belcheri tsingtauense: *the neurula and larva. Acta Anatomica Nipponica 69, 1–13.*

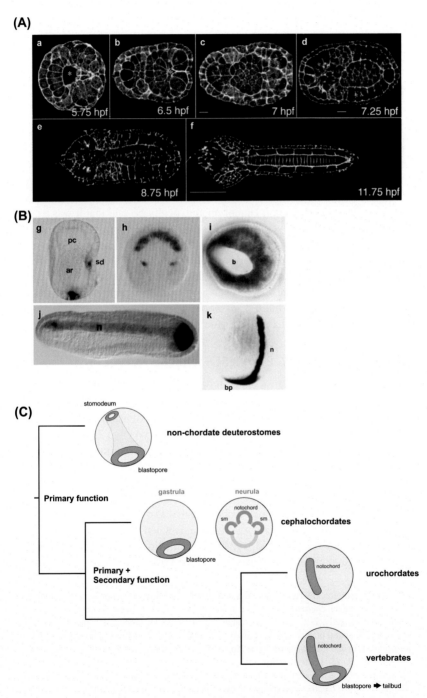

FIGURE 7.2   Notochord formation and *Brachyury* expression. (A) Ascidian notochord development from the onset of gastrulation to the completion of convergent extension. (a) The gastrula, and * indicates the blastopore. (b–d) The neurula. Two rounds of cell division generate 20 (b) and

notochord secretes factors that signal surrounding tissues regarding position and fate (eg, Stemple, 2005; Nibu et al., 2013; Corallo et al., 2015). In this role, the notochord secretes the Sonic Hedgehog signaling molecule to specify cell fates in the ventral CNS region, Wnt signaling molecules to specify pancreatic cell fates, controlling the arterial versus venous identity of the major axial blood vessels, and specifying various cell types in forming somites.

7.5.1 Brachyury: A Key Transcription Factor for Notochord Formation

Because *Brachyury* is one of the key regulatory genes responsible for development of chordate-specific structures, the molecular nature of this gene will be discussed in detail here. In 1990, Herrmann and colleagues succeeded in cloning mouse *Brachyury* via an elaborate chromosomal walk (Wilkinson et al., 1990). *Brachyury* (short tail) or *T* (tail) is named after a mutation in which homozygous embryos die in utero because of deficient mesodermal development, whereas heterozygous embryos are born with short tails (Dobrovolskaia-Zavadskaia, 1927; Chesley, 1935). The gene is expressed in early-stage mesoderm and is subsequently restricted to the notochord. Later, mouse *Brachyury* was shown to act as a tissue-specific transcription factor by specifically binding to a palindrome of 20 bp, the T-site, including 5′-AGGTGTGAAATT-3′ (Kispert et al., 1995). *Brachyury* likely represents the ancestral form of the T-box gene family including *Tbx2*, *Tbx6*, and *Tbr*, with ancient and/or primary functions (reviewed by Papaioannou, 2001, 2014; Showell et al., 2004).

Reflecting its significant role in mouse embryogenesis, *Brachyury* orthologs have been isolated from various metazoans, including frogs, zebrafish (Fig. 7.2B(k)), urochordates (Fig. 7.2B(h)), cephalochordates (Fig. 7.2B(i, j)),

◀ finally 40 notochord cells that form a monolayer epithelium (c). (d–f) Infolding (not shown here) and convergent extension transform the notochord precursor into a column of 40 stacked cells. *Ciona savignyi* embryos are stained with bodipy-phalloidin and imaged with a laser scanning confocal microscope. *Hpf*, hours post fertilization. All images are dorsal view, with anterior to the left. *(From Jiang, D., Smith, W.C., 2007. Ascidian notochord morphogenesis. Developmental Dynamics 236, 1748–1757.)* (B) The expression of *Brachyury* in deuterostomes. (g) Hemichordates: *Bra* is expressed in the blastopore region and stomodeum (sd) invagination region. *ar*, archenteron; *pc*, protocoel. *(From Tagawa, K., Humphreys, T., Satoh, N., 1998. Novel pattern of Brachyury gene expression in hemichordate embryos. Mechanisms of Development 75, 139–143.)* (h) *Ciona*: in primordial notochord cells at the 110-cell stage embryo. *(From Corbo, J.C., Levine, M., Zeller, R.W., 1997. Characterization of a notochord-specific enhancer from the Brachyury promoter region of the ascidian, Ciona intestinalis. Development 124, 589–602.)* (i, j) Amphioxus: on the marginal zone of the blastopore (b) of the gastrula (i), and notochord (n) and tailbud of 18-h embryos (j). *(From Holland, P.W.H., Koschorz, B., Holland, L.Z., Herrmann, B.G., 1995. Conservation of Brachyury (T) genes in amphioxus and vertebrates: Developmental and evolutionary implications. Development 121, 4283–4291.)* (k) *no tail* expression in the blastopore (bp) and notochord of a zebrafish embryo. *(From Schulte-Merker, S., van Eeden, F.J., Halpern, M.E., Kimmel, C.B., Nüsslein-Volhard, C., 1994. no tail (ntl) is the zebrafish homologue of the mouse T (Brachyury) gene. Development 120, 1009–1015.)* (C) A schematic drawing to show the transition from primary gene expression in the blastopore to secondary expression in the notochord.

hemichordates (Fig. 7.2B(g)), echinoderms, protostomes, and diploblasts (see Satoh et al., 2012 for references). As a result, several things became evident. First, only a single copy of *Brachyury* is present in most metazoan genomes. Second, although various transcription factor genes are expressed various times in different tissues and serve multiple functions in embryogenesis, *Brachyury* is expressed only once in the archenteron-invaginating region and the mouth-invaginating region of gastrulae in almost all metazoan taxa (Fig. 7.2C). Third, in sea urchins, functional suppression of *Brachyury* results in the failure of gastrulation (eg, Gross and McClay, 2001). That is, *Brachyury* is transiently expressed in the blastopore region, where it enables cells to invaginate inside of the embryo. Because gastrulation is a morphogenetic event leading to the formation of embryos with two or three germ layers (Stern, 2004; Wolpert and Tickle, 2011), *Brachyury* is essential to embryogenesis of all metazoans; we call this the "primary expression and function" of the gene (Satoh et al., 2012; Fig. 7.2C).

During chordate evolution, *Brachyury* developed an additional or "secondary" expression domain in mesoderm (Fig. 7.2C). As was discussed in the previous section, early cephalochordate embryogenesis offers clues to notochord formation in relation to chordate origins. The notochord is formed by pouching off from the archenteron, and similarly, somites form along both sides of the archenteron (Fig. 7.2C). Interestingly, amphioxus *Brachyury* is expressed in both regions. With its muscle-like properties, the morphogenetic movement of notochord (and somites) in cephalochordate embryos appears as a continuation of the archenteron invagination or secondary invagination. Amphioxus may have recruited secondary *Brachyury* expression for this second invagination-like morphogenetic movement. In this context, it appears that in cephalochordates, the biochemical nature of genes downstream of *Brachyury* is unimportant as long as the components have sufficient stiffness to support tail beating.

On the other hand, in urochordates and gnathostomes the notochord is formed by convergent extension of precursor cells that are bilaterally positioned in the early embryo (Fig. 7.2A). These cells do not possess any muscle-like properties. In vertebrates, *Brachyury* is first expressed in the marginal zone of the blastopore (*Xenopus*) or in the germ ring (*Danio*), subsequently in the notochord (Fig. 7.2B(k) and C), and finally it persists in the tailbud region. *Brachyury* expression in the tailbud region is likely a continuation of marginal zone expression, reflecting an invagination and/or movement of embryonic cells from that region (Gont et al., 1993). In ascidian urochordates, *Brachyury* is expressed exclusively in notochord cells because its expression in embryonic muscle cells is suppressed by *snail* (Fig. 7.4; Yasuo and Satoh, 1993; Corbo et al., 1997; Fujiwara et al., 1998).

7.5.2 Regulatory Networks of *Brachyury*

As discussed previously, a key issue involved in notochord formation from the standpoint of chordate origins is how *Brachyury* acquired its secondary expression and function. In the amphioxus, *Branchiostoma belcheri*, and the frog, *Xenopus tropicalis*, a duplication of *Brachyury* resulted in two tandemly aligned copies of the gene (Holland et al., 1995; Gentsch et al., 2013). It is tempting to speculate that one is the original *Brachyury* with its original function and the other is a new copy with a secondary function, and that the duplication occurred at a very early phase of chordate evolution. At present it is uncertain whether this is so, and this question should be investigated. However, urochordates and most vertebrates have only one copy of *Brachyury*. This makes studies of the gene regulatory network of *Brachyury* (Br-GRN) more complicated.

To date, extensive studies have been performed to characterize Br-GRN (reviewed by Satoh et al., 2012). Difficulties of Br-GRN studies include the following: that (1) experiments to discover *cis*-regulatory modules in acorn worms (hemichordates) and lancelets (cephalochordates) have failed for technical reasons, (2) discrimination of *cis*-regulatory modules for primary and secondary expression of each is technically complicated. This implies that primary expression is transient in archenteron-invaginating cells whereas secondary expression is constitutive in notochord-forming cells, and (3) Br-GRNs among five deuterostome taxa are not always comparable because of the complexity of Br-GRN itself in each taxon (Fig. 7.3).

Nevertheless, Br-GRN has been fully revealed in the ascidian, *Ciona intestinalis* (Fig. 7.3A and B) because *Ci-Bra* is expressed exclusively in 40 notochord cells, the embryonic lineages of which have been completely described (Satoh, 2014). The upstream genetic cascade leading to initiation of *Brachyury* (*Ci-Bra*) expression involves maternally expressed β-catenin, which activates zygotic expression of *FoxD*, *FoxA*, and *FGF8/16/19* (Fig. 7.3A). *FoxD* is a direct target of β-catenin and is expressed in presumptive notochord blastomeres at the 8- and 16-cell stages. FoxD activates *ZicL* at the 32-cell stage, and ZicL activates *Bra* at the 64-cell stage. *FoxA* and *FGF8/16/19* are also involved in *Bra* transcription activation, and maternal P53 activates *ZicL* at the 32-cell stage. In addition, *Ci-Bra* has an autonomous, suppressive regulation loop for its expression (Fig. 7.3A).

In addition, approximately 400 genes are identified as *Ci-Bra* downstream. These genes function in gastrulation movements, formation of the notochord, and signaling from the notochord to neighboring cells, as previously mentioned (Fig. 7.3A; Takahashi et al., 1999; Hotta et al., 2008). Some *Ci-Bra* downstream genes are well conserved in vertebrates and are likely to participate in formation and function of notochord cells (José-Edwards et al., 2011). On the other hand, almost no correspondence was found in *Bra*-downstream genes between sea urchins (Rast et al., 2002) and ascidians (Hotta et al., 2008), presumably because the former performs its primary function whereas the latter has assumed a secondary function. In vertebrates, including *Xenopus* and zebrafish, Br-GRN function seems more complex (Fig. 7.3C). This is because notochord development is profoundly associated with establishment of the embryonic anteroposterior (AP) and DV axes. The canonical Wnt/β-catenin pathway, the Fox

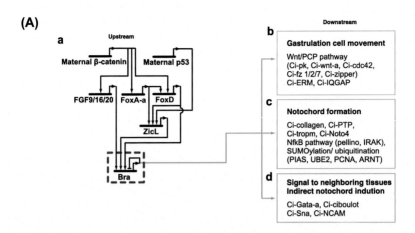

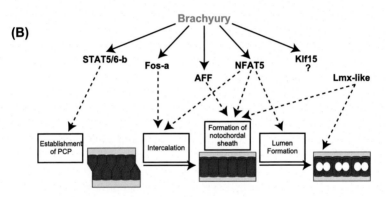

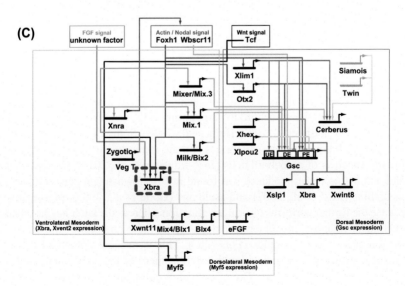

transcription factors, the fibroblast growth factor (FGF) signaling pathway, and the transforming growth factor-β (BMP/Nodal) signaling pathway are involved in Br-GRN. Recent investigations of *Brachyury*-mediated developmental pathways using more sophisticated techniques, such as Chip-seq, reveal very complex Br-GRN in vertebrates (eg, Morley et al., 2009; Evans et al., 2012; Gentsch et al., 2013; Lolas et al., 2014).

Fig. 7.4 shows *cis*-regulatory modules of the 5′ upstream sequences of *Brachuyry* in the ascidian *Halocynthia roretzi* (Fig. 7.4A; Matsumoto et al., 2007), *C. intestinalis* (Fig. 7.4B; Corbo et al., 1997), and the frog *Xenopus laevis* (Fig. 7.4C; Koide et al., 2005), respectively. As far as they were explored, the 5′ upstream sequences contain domains that act as positive or negative control, including the Ets binding sites for the FGF signaling pathway (Fig. 7.4A), Zic and Fox binding sites for positive control (Fig. 7.4A and B), and Sna binding sites for negative control (Fig. 7.4B). The *Xenopus* 5′ upstream sequences appear more complex than those of ascidians because of incorporation of additional regulatory factors (Fig. 7.4C). However, further research is required to characterize the *cis*-modules of chordate *Brachyury*.

7.6 SOMITES (MYOTOMES): A MESODERMAL NOVELTY

A comparison of muscle development in acorn worms and lancelets might highlight considerable differences between them because of their different lifestyles (eg, Diogo and Zimmerman, 2015). In acorn worms, a major muscle structure originates from the procoele, the most anterior part of the embryo (Fig. 1.3B). This region of embryonic mesoderm forms the proboscis of enteropneusts, which is muscular and is used for digging and feeding. Muscle cells for undulating locomotion of the worm are also present in the wall of the trunk. On the other hand, in lancelets, bilateral pouching off from the archenteron forms most

◀ FIGURE 7.3 **Gene regulatory networks of *Brachyury*.** (A) Gene regulatory networks upstream and downstream of *Brachyury* in embryos of the ascidian, *C. intestinalis*. Upstream positive regulators include maternal *β-catenin* and *P53*, and zygotic *FoxD*, *FoxA*, *FGF9/16/20*, and *ZicL*. Ci-Bra directly activates *Ci-tropm* and other notochord-specific genes to induce notochord differentiation. Cell polarity-related genes, such as *Ci-pk* and *Ci-Fz1/2/7*, are involved in convergent extension movements during gastrulation. These signaling components are induced directly or indirectly by *Ci-Bra*. *(Modified from Hotta, K., Takahashi, H., Satoh, N., Gojobori, T., 2008. Brachyury-downstream gene sets in a chordate,* Ciona intestinalis: *integrating notochord specification, morphogenesis and chordate evolution. Evolution & Development 10, 37–51.)* (B) Notochord transcription factors, their relationship with *Ci-Bra (solid arrows)* and their putative functions in notochord development *(dotted arrows)*. Notochord cells are depicted in *red* whereas the notochordal sheath is shown in *purple*; *white circles* represent intercellular lumens. *(From José-Edwards, D.S., Kerner, P., Kugler, J.E., Deng, W., Jiang, D., Di Gregorio, A., 2011. The identification of transcription factors expressed in the notochord of* Ciona intestinalis *adds new potential players to the brachyury gene regulatory network. Developmental Dynamics 240, 1793–1805.)* (C) Gene regulatory networks upstream and downstream of *Brachyury* in embryos of the frog, *X. laevis*. *(From Koide, T., Hayata, T., Cho, K.W.Y., 2005.* Xenopus *as a model system to study transcriptional regulatory networks. Proceedings of the National Academy of Sciences of the United States of America 102, 4943–4948.)*

(A)

Halocynthia roretzi

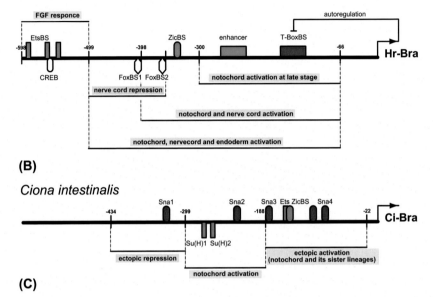

(B)

Ciona intestinalis

(C)

Xenopus laevis

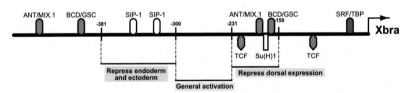

FIGURE 7.4   A schematic representation of *cis*-regulatory motifs in the 5′-upstream region of *Brachyury*. (A) *Hr-Bra* of the ascidian *H. roretzi*. (*From Matsumoto, J., Kumano, G., Nishida, H., 2007. Direct activation by Ets and Zic is required for initial expression of the* Brachyury *gene in the ascidian notochord. Developmental Biology 306, 870–882.*) Binding motifs are shown such as EstBS for Ets binding sites, FoxBS for Fox binding sites, ZicBS for Zic binding site, and T-BoxBS for Bra binding site, respectively. (B) *Ci-Bra* of *C. intestinalis*. (*From Corbo, J.C., Levine, M., Zeller, R.W., 1997. Characterization of a notochord-specific enhancer from the* Brachyury *promoter region of the ascidian,* Ciona intestinalis. *Development 124, 589–602.*) Sna, Snail binding site and Su(H), Suppressor of hairless binding site. (C) *Xbra* of *X. laevis*. ANT/MIX1 for homeodomain transcription factor biding sites, BCD/GSC for another homeodomain transcription factor binding site. TCF for TCF binding site, and SIP-1 for Smad interacting protein-1 binding site, respectively. (*From Koide, T., Hayata, T., Cho, K.W.Y., 2005.* Xenopus *as a model system to study transcriptional regulatory networks. Proceedings of the National Academy of Sciences of the United States of America 102, 4943–4948.*)

of the somites, and posterior somites pinch off sequentially from the tailbud (Conklin, 1932; Gomez et al., 2008; Mansfield et al., 2015). Because of the continuity of somite musculature of the larval tail with the segmented myotome of the adult body wall, the structure is traditionally called "body-wall muscle." As discussed previously, *Brachyury* is expressed in the bilateral mesoderm invagination region of amphioxus embryos (Fig. 7.2C). In addition, in the frog, *X. laevis*, *Brachyury* plays a role in somite formation (Fig. 7.6D).

In vertebrates, a "clock and wave-front" mechanism, involving oscillating Notch and Wnt signaling and posterior FGF signaling gradients, divides the paraxial mesoderm into a series of somites (eg, Hubaud and Pourquié, 2014). Despite these mechanistic differences, amphioxus displays vertebrate-like segmental expression of Notch and Wnt signaling components in nascent somites and requires FGF signaling for forming and maturing anterior and posterior somites (Beaster-Jones et al., 2008). Thus somitogenesis in all living chordates, and presumably their last common ancestors, involved iterated Notch-Delta and Wnt signaling and FGFs.

Remodeling of muscle structures during chordate evolution was likely accompanied by genetic changes, as in the case of actin isoforms. Vandekerckhove and Weber (1979, 1984) found that mammalian skeletal muscle actins may be distinguished from cytoplasmic actins at 25 or more positions in the amino acid sequences (Fig. 7.5). In addition, they also reported that vertebrate muscle actins differ from invertebrate muscle actins (including those of sea stars), which show vertebrate cytoplasmic actin properties (Vandekerckhove and Weber, 1984). Later, it was shown that vertebrate muscle-type actins occur in urochordates and cephalochordates (Kusakabe et al., 1992, 1997; Chiba et al., 2003). Fig. 7.5 compares amino acid compositions of actins from vertebrate muscle and cytoplasm, ascidian muscle and cytoplasm, and amphioxus muscle. In each taxon an ancestral actin gene has duplicated to produce many isoforms. It is evident that (1) in vertebrates, muscle actins are different in amino acid composition from cytoplasmic actins; (2) similarly, in ascidians, muscle actins, which somewhat resemble vertebrate muscle actins, are distinguishable from their corresponding cytoplasmic actin isoforms; and (3) amphioxus actins appear intermediate between vertebrate muscle and cytoplasmic actins. On the other hand, hemichordate actins resemble the cytoplasmic actins of vertebrates. This pattern is also seen in other muscle proteins, including myosin heavy chains and myosin regulatory light chains. These results suggest the occurrence of genetic changes of muscle-related genes during chordate evolution—first in the diversification of cephalochordates from the deuterostome lineage and second in the diversification of the Olfactores from the cephalochordate lineage. Details of these changes should be explored in future studies.

7.7 THE POSTANAL TAIL: A MESODERMAL NOVELTY

The postanal tail is the structure or region posterior to the anus (Fig. 1.4). Tail beating provides a major source of locomotive power. The tail of chordate larvae and juveniles consists of three germ layers, as does the trunk region. Recent studies in vertebrates focused on how the postanal tail is formed and what kind of transcription factors and signaling pathway molecules are involved in its

Homo sapiens muscle type

Hs-skeletal_muscle	S	T	C	L	V	I	T	V	L	N	M	V	N	T	I	A	N	A
Hs-cardiac_muscle	T	T	C	L	V	I	T	V	L	N	M	V	N	T	I	A	N	A
Hs-smooth_muscle	T	T	C	L	C	I	T	V	L	N	M	V	N	T	I	A	N	A
Hs-aorta_muscle	T	T	C	L	C	I	T	V	L	N	M	V	N	T	I	A	N	A

Homo sapiens cytoplasmic type

Hs-cytoplasmic_beta	I	A	V	M	C	V	V	T	M	T	L	T	Q	A	L	C	T	S
Hs-cytoplasmic_gamma	I	A	I	M	C	V	V	T	M	T	L	T	Q	A	L	C	T	S

Ciona intestinalis muscle type

Ci_MA1	Q	T	C	L	V	I	T	V	M	N	A	V	Q	T	I	A	N	A
Ci_MA2	Q	T	C	L	V	I	V	V	L	N	M	V	Q	T	I	S	N	A

Ciona intestinalis cytoplasmic type

Ci_CA1	V	A	V	M	C	V	V	T	L	T	I	T	Q	T	I	A	T	S
Ci_CA6	V	A	V	M	C	V	V	T	F	T	R	T	K	A	L	A	T	S

Branchiostoma floridae

Bfl_Ac1	T	T	V	L	V	V	T	V	L	T	M	V	Q	A	L	T	N	S
Bfl_Ac2	T	T	C	L	C	V	C	V	L	T	M	V	Q	A	L	T	N	S
Bfl_Ac3	V	A	V	M	C	I	V	T	L	T	M	V	Q	T	I	A	T	S
Bfl_Ac4	V	A	V	M	C	V	V	S	L	T	L	T	Q	S	L	T	T	S

Ptychodera flava

Pfl_Ac2	I	A	V	M	C	V	V	T	L	V	Q	T	G	A	L	A	T	C
Pfl_Ac3	V	A	V	M	C	V	V	T	M	T	L	T	Q	A	L	S	T	S

Saccoglossus kowalevskii

Skw_Ac2	V	A	V	M	C	V	V	T	M	T	L	T	Q	A	L	P	T	S
Skw_Ac3	V	A	V	M	C	V	V	V	L	V	E	T	E	C	L	G	T	S

FIGURE 7.5 Comparison of amino acid composition of actin isoforms in deuterostomes. Amino acid residues of vertebrate muscle actin proteins are shown in *black boxes* and those of vertebrate cytoplasmic actin proteins with *black letters* in *white boxes*. Amino acid residues that differ between muscle and cytoplasmic actin proteins are in *gray boxes*. *Modified from Chiba, S., Awazu, S., Itoh, M., Chin-Bow, S.T., Satoh, N., Satou, Y., Hastings, K. E.M., 2003. A genomewide survey of developmentally relevant genes in* Ciona intestinalis *– IX. Genes for muscle structural proteins. Development Genes and Evolution 213, 291–302.* Courtesy of Fuki Gyoja for analyses of *B. floridae, P. flava and S. kowalevskii.*

formation (eg, Beck, 2015). Because this morphogenesis involves reorganization of ectoderm, mesoderm, and endoderm, the cellular and molecular mechanisms are proving to be more complex than we thought. However, it is highly likely that the postanal tail is produced by morphogenetic cellular migrations associated with tailbud formation (eg, Beck, 2015). The tailbud functions as a part of an organizer, presumably a continuation of the dorsal lip of vertebrate embryos, for tail elongation in chordates (Gont et al., 1993). One of the features of *Brachyury* expression in chordate embryos (especially in vertebrate) is that *Brachyury* expression in the blastopore marginal zone continues to the tailbud region after most of the endomesodermal cells have migrated inside of the embryos (Fig. 7.2C; Gont et al., 1993). The function of *Brachyury* in the tailbud has not yet been well characterized, and this is one of the urgent questions in relation to the origin of chordates. An intriguing fact is that *Brachyury* is expressed at the tip of the anus of enteropneust hemichordates, suggesting that the anus retains some characteristics of other cell invaginations.

In summary, the origin of chordates was accomplished by novelties in mesoderm formation. At least the notochord, the somites, and the tailbud are not seen in nonchordate deuterostomes. Intriguingly, one T-box gene, *Brachyury*, is likely to play pivotal roles in the development of these novelties, suggesting its significance in chordate origins.

7.8 THE DORSAL CENTRAL NERVOUS SYSTEM: AN ECTODERMAL NOVELTY

Formation of the dorsal CNS in chordates is a conspicuous developmental event that does not occur in nonchordate deuterostomes (eg, Heasman, 2006; Holland, 2009; Angerer et al., 2011). In the cephalochordate amphioxus, soon after gastrulation, typical and dynamic neurulation movement occurs on the dorsal side of the embryo. This morphogenetic movement appears as a distinct event, taking place earlier than that of notochord formation beneath it. In urochordate embryos, a well-organized neural plate is formed on the dorsal side of the embryo, followed by curling and fusion of both edges of the plate to form the neural tube. In vertebrates, as seen in *Xenopus* embryos, neurulation is distinct. The significance of neural tube (not nerve cord) formation in chordate evolution was discussed in Section 7.1.

7.8.1 The Gene Regulatory Network for Neural Development

CNS fate is determined from embryonic ectoderm and is separated during early embryogenesis from epidermal fate. Because CNS development has been the focus of vigorous neuroscience investigation, here only a brief synopsis of the GRNs involved in epidermal versus neural specification in *Xenopus* embryos will be mentioned (Heasman, 2006; Holland, 2015).

At the mid-blastula transition throughout the mid-blastula of *Xenopus* embryos, the bone morphogenetic protein (BMP) signaling pathway is initially activated, downstream of maternal BMP2 and BMP7 activity, except in the embryonic dorsal animal quadrant (Fig. 7.6A; Faure et al., 2000; Schohl and Fagotto, 2002). The silencing of BMP signaling in this quadrant may be

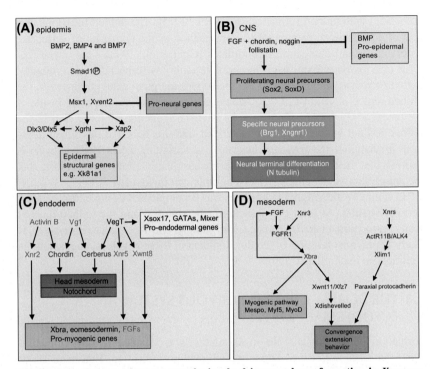

FIGURE 7.6 Gene regulatory networks involved in germ layer formation in *Xenopus*.
(A) The epidermal regulatory network. Bone morphogenetic protein (BMP) signaling directly activates target genes, including *Xvent2* and *Msx1* (which activate the epidermal pathway and suppress neural fates when ectopically overexpressed). Xvent2 and Msx1 regulate the more restricted pro-epidermal genes *Xap2*, *Dlx3*, and *Xgrhl1*, which directly regulate epidermal structural genes such as the cytokeratin gene *Xk81.1a*, but that lack neural inhibiting functions. (B) The neural regulatory network. Animal cells are directed toward a neural fate beginning at the late blastula stage and continuing through gastrulation. Neural induction requires both suppression of the BMP-Smad1-Msx pathway by the BMP antagonists chordin, noggin, and follistatin and the activation of fibroblast growth factor (FGF) signaling. Cells begin to express proneural genes (*Sox2* and *SoxD*) at the gastrula stage. By the end of gastrulation, the deeper layer of the proneural ectoderm activates specific neuronal precursor markers, including *Xngnr1*, which in turn regulate primary neuronal differentiation via *XneuroD* and *N-tubulin*. Primary neurogenesis requires the maternal chromatin remodeling protein Brg1. (C) The endodermal regulatory network. VegT activates transcription of proendodermal transcription factors, including Xsox17, GATA5, and Mixer, and of signaling molecules, including Xnr5, Wnt8, and the signaling antagonist Cerberus. Vg1 and activin B also activate signaling pathways that lead to transcription of the signaling antagonists chordin and Cerberus. The signaling molecules Xnr2, Xnr2, Xnr5, and Wnt8 are important mesoderm-inducing molecules, responsible for transcriptional activation of the promyogenic genes, *Xbra* and *eomesodermin*, and *FGF* genes, which can regulate specific myogenic genes such as *MyoD*. Cerberus suppresses Xnr and Wnt, and BMP signals extracellularly and is required for head (prechordal) mesoderm formation, whereas BMP suppression by chordin is required for notochord formation. Signaling molecules are shown in *blue* and signaling antagonists in *red*. (D) The role of Brachyury. FGF activates pathways through *FGFR1* and *Xbra*, leading to convergence extension movements and myogenic specification. Xnr3 also activates FGF receptor-dependent *Xbra* expression, but only in the embryonic dorsal area. Xbra causes convergence extension move-

the result of the early expression of the BMP antagonists, noggin or chordin (Kuroda et al., 2004). Animal cap regions explanted from mid-blastulae follow an epidermal differentiation pathway that is dictated by BMP signaling (Fig. 7.6A). The epidermal regulatory network downstream of BMP signaling includes the transcriptional activators *Xvent2* (Onichtchouk et al., 1996) and *Msx1* (Suzuki et al., 1997), which activate the epidermal pathway and suppress proneural genes when ectopically overexpressed (Fig. 7.6A). In turn, these genes activate more restricted proepidermal genes, including *Dlv3/Dlx5*, *Xgrh1*, and *Xap2*, which can directly regulate epidermal structural genes but do not have neural repressive roles (Tao et al., 2005).

When all BMPs and the dorsally expressed BMP-like molecule, antidorsalizing morphogenetic protein, are depleted, the entire outer layer of the embryo expresses neural markers (Reversade and De Robertis, 2005). This result led to the idea that neural fate is the "default state," such that if cells receive no specific signal, they otherwise select neural fate. However, it has recently been shown that cell dissociation actually activates FGF signaling and inhibits Smad1 via mitogen-activated protein kinase phosphorylation of its linker region (Kuroda et al., 2005). Later, in vivo studies suggest that expression of the proneural genes, *Sox2* and neural cell-adhesion molecule gene, *NCAM*, depends on low levels of FGF signaling at the blastula stage, independent of BMP antagonism (Fig. 7.6B; Delaune et al., 2005). Namely, FGF (+chordin, noggin, and follistatin) activates the proneural genes, *Sox2* and *SoxD*, which activate specific neural precursor genes, *Brg1* and *Xngnr1*. Thus neural specification may require FGF signaling and BMP antagonism whereas epidermal fate is determined by BMP signaling (Fig. 7.6B).

The nervous system of nonchordate deuterostome larvae consists of the apical organ and ciliary bands (Bishop and Burke, 2007; Nielsen, 2012). Because the apical organ degenerates at metamorphosis and the new nervous centers take over, the apical organ does not necessarily need to be discussed in relation to the chordate CNS. Another structure, the ciliary band of dipleurula (auricularia) larvae, has also been a topic of discussion in relation to the origin of the chordate CNS. As mentioned previously, Garstang (1928a) proposed that during chordate evolution the ciliary band of dipleurula larvae moved dorsally, finally fusing mid-dorsally, and enclosing the entire aboral surface of the larvae (Fig. 2.3). It should be emphasized here that morphogenesis to form the dorsal neural tube in chordate embryos is completely different from that which forms the ciliary band nervous systems of ambulacrarian embryos.

◀ ments by activating the expression of zygotic *Wnt11*. Wnt11 signaling causes convergence extension in a *dishevelled*-dependent manner that does not involve β-catenin, the so-called "noncanonical Wnt pathway." A separate pathway regulating convergence extension involves the activation of paraxial protocadherin downstream of Xlim1. The cytoplasmic kinase regulator genes *sprout* and *spred* determine whether cells undergo convergence extension or somite formation in response to FGF receptor stimulation. *From Heasman, J., 2006. Patterning the early* Xenopus *embryo. Development 133, 1205–1217.*

7.8.2 Regionalization of the Central Nervous System

After its formation, the neural tube becomes compartmentalized with different functions along the AP and DV axes. Extensive comparative analyses of neuro-anatomy and developmental gene expression between vertebrates and cephalo-chordates have demonstrated that lancelets lack a neural crest, most placodes, and a midbrain–hindbrain (MHB) organizer, suggesting that these are vertebrate innovations (Holland, 2003, 2009; see Chapter 9). The anterior expanded portion of the amphioxus CNS is probably equivalent to a diencephalic forebrain plus a small midbrain, whereas the remainder of the CNS is homologous to the vertebrate hindbrain and spinal cord (Fig. 7.7). This suggests that vertebrate features such as the neural crest and MHB were built on a foundation already present in the ancestral chordate (Holland, 2009).

Regionalization of the chordate neural tube is regulated as an aspect of AP and DV axis formation of the entire body. Along the AP axis, Wnt and reti-noic acid signaling probably act on Hox genes and other transcription factor genes to establish regional identities of AP domains of the body axis, includ-ing the boundary between the foregut and hindgut and the main subdivisions of the CNS (Fig. 7.7; eg, Mizutani and Bier, 2008). Expression domains of transcription factors and signaling molecules, including *Otx*, *Gbx*, *Fgf 8/17/18*, *Wnt1*, *En*, *Pax2/5/8*, *Irx*, and *Hox1-4*, along the AP axis of the CNS, are also mostly conserved across chordates and presumably reflect expression domains of the chordate ancestor. Precisely how this patterning was generated is less clear because current data suggest that neither amphioxus nor tunicates have unambiguous, functionally validated homologs of two vertebrate CNS signaling centers—the isthmic organizer and the zona limitans intrathalamica.

In vertebrates and cephalochordates the DV axis is established by opposing Nodal and BMP signaling gradients, with Chordin-mediated BMP inhibition in the dorsal ectoderm, segregating the presumptive CNS from epidermal (or gen-eral) ectoderm (eg, Baker, 2008). Along the DV axis of the CNS, all chordates have a molecularly distinct dorsal domain that expresses *pax3/7*, *msx*, and *zic* genes and generates sensory interneuron cells, a ventral floor plate express-ing hedgehog ligands, and an intervening bilateral domain flanking the neural tube lumen and generating motor and visceral neurons. However, many ques-tions remain to be answered to understand the molecular and cellular mecha-nisms involved in chordate CNS evolution, especially neural tube formation and regionalization involving different functions.

7.9 HATSCHEK'S PIT: AN ECTODERMAL NOVELTY

Comparison of ectoderm-derived structures between ambulacrarians and chor-dates makes it clear that at least three major innovations occurred during chor-date evolution: the formation of a dorsal neural tube, a neurogenic sensory placode, and a neural crest. The latter two appear not to be innovations shared

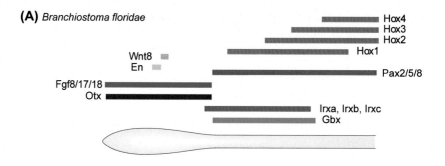

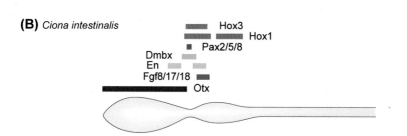

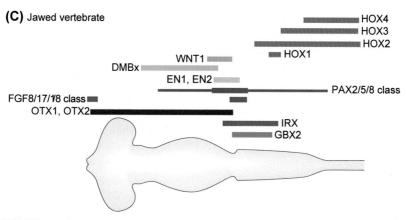

FIGURE 7.7   Comparison of midbrain–hindbrain gene expression. (A) Amphioxus, (B) tunicates, and (C) a generic jawed vertebrate. Domains of gene expression are shown above dorsal views of the brain. Where the expression domain of a particular gene differs among vertebrates, that of the chick has been given. In the ascidian tunicate, *C. intestinalis* (B), *Hox2* and *Hox4* are not expressed in the central nervous system (CNS), and *Wnt1* and *Gbx* have been lost. Expression of *Irx* is unknown. The tail nerve cord has no nerve cell bodies and the tail is resorbed at metamorphosis. In the amphioxus, *Branchiostoma floridae* (A), unlike vertebrates (C), neither *Dmbx* nor *Wnt1* is expressed in the CNS. *From Holland, L.Z., 2009. Chordate roots of the vertebrate nervous system: expanding the molecular toolkit. Nature Reviews Neuroscience 10, 736–746.*

by all three chordate taxa but are specific to vertebrates. However, the sudden appearance of the neural crest and sensory placode in the early branches of vertebrates has puzzled biologists for more than a century. Are the neural crest and the sensory placode vertebrate innovations or are their homologs in cephalochordates and urochordates? The neural crest and neural placode will be discussed in detail in Sections 10.1.1 and 10.1.2, respectively. Here, only the story of the cephalochordate Hatschek's pit is related.

The sensory placode is a thickening of a specific region of epidermis in developing vertebrate larvae (Section 10.1.2). There are at least eight types of placode (see Fig. 10.2). Cells originating from different regions develop into various cell types that help form sensory organs. It has been thought that most placodes are vertebrate novelties, but the exception to this may be Hatschek's pit (Fig. 1.4B). Hatschek's pit is formed from a preoral pit and is identifiable as an evagination of the roof of the pharynx that makes contact with a small area at the base of the brain, and as such it appears similar to the juxtaposition of the anterior and posterior lobes of the pituitary in the vertebrate brain. Hatschek's pit expresses the vertebrate pituitary gene *Pitx*, reinforcing the homology of these structures suggested by anatomical (adult anatomy, relative position, and embryological origin) and functional evidence. Therefore it is likely that the cephalochordate Hatschek's pit is an ectodermal novelty of chordates.

7.10 THE ENDOSTYLE: AN ENDODERMAL NOVELTY

The endostyle is a longitudinal, ciliated, grooved organ on the ventral wall of the pharynx that secretes mucoproteins into the alimentary canal for filter feeding (Fig. 7.8). Because the organ is not found in hemichordates and echinoderms, but is formed in cephalochordates, urochordates, and larval lampreys, it is thought to have arisen as an innovation of chordates. In *Ciona*, the adult endostyle is derived from larval endoderm. In lampreys, the endostyle is developed near the pharynx, and some component cells change into follicular epithelial cells of the thyroid gland. In fishes and amphibians, an endostyle-like organ is never seen during embryogenesis, and the thyroid is formed in the juvenile stage. Because the endostyle absorbs iodine, an evolutionary link has been suggested between this organ and the follicular thyroid of vertebrates. This notion is supported by the expression of *TTF-1* (*Nkx2.1*: see Section 6.2) in the endostyle of *Ciona*, amphioxus, and lamprey. *TTF-1* is a transcription factor gene that controls survival of thyroid follicular cells at the beginning of organogenesis and regulates the expression of thyroid-specific structural genes in adult vertebrates.

The expression profile of toolkit genes in the endostyle has been well characterized in lancelets, ascidians, and larvaceans (Fig. 7.8) (eg, Hiruta et al., 2005). The *Ciona* endostyle consists of nine zones, including supporting zones (zones 1, 3, and 5), glandular zones (zones 2, 4, and 6), and iodine-binding zones (zones 7, 8, and 9; Fig. 7.8B(c)), whereas the amphioxus endostyle contains seven functional compartments, such as supporting (zones 1 and 3), glandular

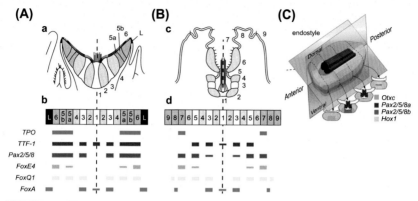

FIGURE 7.8 The endostyle of cephalochordates and urochordates. (A, B) Comparison of histological components and gene expression patterns in the endostyle of amphioxus (a, b) and *Ciona* (c, d). Transverse sections. (a, c) Both endostyles possess supporting elements, glandular elements, and thyroid-equivalent elements. Supporting elements, glandular elements, thyroid-equivalent elements, and the region lateral to zone 6 of the amphioxus endostyle are indicated in *white, lightly shaded, darkly shaded,* and *black boxes,* respectively. The midlines of the endostyle are indicated by *dotted lines.* (b, d) Gene expression pattern. Each component is aligned with an equal interval and is indicated by numbers for each zone. A distinct expression domain (*bold bar*) and a weak and/or occasional expression domain (*thin bar*) are shown below the corresponding zones. *(From Hiruta, J., Mazet, F., Yasui, K., Zhang, P.J., Ogasawara, M., 2005. Comparative expression analysis of transcription factor genes in the endostyle of invertebrate chordates. Developmental Dynamics 233, 1031–1037.)* (C) Schematic three-dimensional reconstruction of the larvacean endostyle summarizing anteroposterior regionalization revealed by the expression domains of *Otxc* (green), *Pax2/5/8a* (red), *Pax2/5/8b* (blue), and *Hox1* (yellow) and showing virtual representations of cross-sections at different levels of the anteroposterior axis. *(From Cañestro, C., Bassham, S., Postlethwait, J.H., 2008. Evolution of the thyroid: Anterior-posterior regionalization of the* Oikopleura *endostyle revealed by Otx, Pax2/5/8, and Hox1 expression. Developmental Dynamics 237, 1490–1499.)*

(zones 2 and 4), and iodine-binding zones (zones 5a, 5b, and 6; Fig. 7.8A(a)). In *Ciona*, not only structural genes but also transcription factor genes are specifically expressed in each zone, suggesting involvement of these genes in regionalization of the endostyle. For example, *TTF-1* is expressed in zones 1, 3, and 5. *TPO* occurs in zone 7. *Pax2/5/8* is expressed in zones 3, 5, and 7. *FoxE4* is found in zones 5 and 7. *FoxQ1* is localized in zones 3, 5, 7, and 8 whereas *FoxA* is restricted to zones 3 and 8 (Fig. 7.8B(d)). The amphioxus endostyle shows a similar transcription factor gene expression profile along the basal-proximal axis of the *Ciona* pharynx (Fig. 7.8A(b)), suggesting that the genetic basis of endostyle function was already in place before separation of the two lineages. In addition, transcription factor gene expression associated with anterior-posterior regionalization of the endostyle is seen in the larvacean *Oikopleura dioica* (Cañestro et al., 2008). *Otx* is expressed rostrally, *Hox1* caudally, and two *Pax2/5/8* paralogs centrally (Fig. 7.8C). Because the ordered expression of *Otx, Pax2/5/8,* and *Hox1* displays patterning in the endoderm-derived endostyle and the ectoderm-derived CNS, this gene set is highly likely to comprise a developmental genetic toolkit of stem bilaterians, which repeatedly provides AP positional information in various developmental events.

Interestingly, the endostyle is a component of the digestive system whereas the thyroid pertains to the endocrine system. The reason for this transition is that the earlier clade of vertebrates changed its feeding mode. Whereas the endostyle is a chordate innovation, cephalochordates and urochordates are highly specialized filter feeders in comparison to acorn worm hemichordates, which are detritus feeders. During vertebrate evolution, the method of food capture changed from filter feeding to item capture. In this transition, ancestral vertebrates did not abandon the endostyle but used it to create the thyroid. Two sets of toolkit genes were incorporated during the innovation of this organ: one for organization of components along the basal-proximal axis and the other along the AP axis. Recruitment of these toolkits in the process of endostyle formation illuminates one way that they have been co-opted. This raises the question of how these toolkits were used during the transition from endostyle to thyroid. Although no reports have shown their persistence in formation of the thyroid, if this is the case, then silencing of the endostyle-specific toolkits may have been required for the transition. That is, although these toolkit genes continue to be expressed, the endostyle persists throughout the life of the individual (for example, of lampreys).

7.11 CONCLUSIONS

One of the important questions is whether chordate features developed by a single modification of GRNs or a complex mixture of independent modifications. As discussed in this chapter, at present it seems unlikely that chordate features emerged by controlling a single network of regulatory genes, rather independently in different germ layers and different structures, although *Brachyury* played pivotal roles in mesoderm-derived structures specific to chordates. I think that these suites of regulatory genes are key players driving chordate origins; therefore I called them each as new organizers. The term *organizers* as used here connotes the same function as in Spemann's organizer or the (mesoderm) organizing center discussed in Chapter 8.

Simultaneously, the term might be used to represent mechanisms that are involved in the emergence of structures that characterize chordates. The new organizers hypothesis is similar to those discussed in Chapters 6, 8, and 9.

Chapter 8

The Dorsoventral-Axis Inversion Hypothesis: The Embryogenetic Basis for the Appearance of Chordates

The dorsoventral (DV)-axis inversion hypothesis, currently the most favored explanation of chordate origins, has received strong support from molecular developmental biology. However, the inversion hypothesis cannot explain the occurrence of chordate-specific structures. On the other hand, the new organizers hypothesis does not explain how the DV-axis inverted nor how the inversion relates to development of chordate-specific characters. This chapter will discuss the embryogenetic basis of chordate origins. Can we bridge the gap between the two hypotheses?

In Chapter 7, I discussed the characteristic features of chordates, including the notochord, the somites, and the dorsal neural tube. I have proposed the new organizers hypothesis to explain how these features developed to promote the appearance of chordates, although many details remain to be explained. Because the chordate-specific structures develop during embryonic and/or larval stages, more attention needs to be paid to the embryological basis for chordate-specific innovations. Currently the most favored or most widely accepted hypothesis to explain chordate origins is the dorsoventral (DV)-axis inversion theory (Arendt and Nübler-Jung, 1994; De Robertis, 2008; Mizutani and Bier, 2008; Holland et al., 2015; Lowe et al., 2015). Although the inversion hypothesis has received strong support from molecular developmental biology, it cannot explain chordate structural novelties. On the other hand, although the new organizers hypothesis may explain chordate novelties, it does not answer the question of how the DV-axis inversion occurred. That is, further discussion to meld the inversion hypothesis with the organizers hypothesis is essential to resolve the question of chordate origins. We have to understand molecular developmental mechanisms underlying embryogenesis in ambulacrarians and chordates.

Therefore this chapter discusses the embryological basis for chordate body plan formation. Embryologically, as in other bilaterians, the deuterostome body plan involves the formation of four embryonic axes: the animal-vegetal (An–Vg) axis, the anteroposterior (AP) axis, the DV axis, and the left-right (LR) axis. Receiving signals or positional cues along the four axes, the three germ layers—ectoderm,

Chordate Origins and Evolution. http://dx.doi.org/10.1016/B978-0-12-802996-1.00008-0

endoderm, and mesoderm—are formed, from which various organs and tissues are developed. I will begin discussion of chordate embryogenesis using amphibian embryos as an example because amphibian embryogenesis has been most intensively studied, giving rise to such concepts as Spemann's organizer, the Nieuwkoop center, and the three-signal model (Agius et al., 2000; Heasman, 2006; Gilbert, 2013; Wolpert and Tickle, 2011). Then AP- and DV-axis formation in early deuterostome embryos will be discussed in an attempt to bridge the gap between the inversion and organizers hypotheses (recent reviews by Lowe et al., 2015; Holland et al., 2015; Kiecker et al., 2015).

The discovery of conserved, pan-bilaterian mechanisms for the development of the AP and DV body axes has advanced or changed our interpretation of the embryonic basis of animal evolution. Early axiation processes of embryos are robust, using conserved suites of genes that are responsible for establishing basic regional differences in cells along all bilaterian embryonic axes and reflecting an extensive genetic regulatory network spread across the developing embryo. The resulting map of conserved expression domains reveals clear relationships between disparate body plans and provides a window into the organization of expression domains in the deuterostome ancestor, as will be discussed later. Early axiation processes are separate from later processes of morphogenesis, organogenesis, and cell differentiation, in which different taxa form diverse organs and tissues.

8.1 SPEMANN'S ORGANIZER, THE NIEUWKOOP CENTER, AND THE THREE-SIGNAL MODEL

Cellular and molecular mechanisms involved in the formation of basic chordate body plans have been well characterized in amphibian embryos (Fig. 8.1A) (Gilbert, 2013; Wolpert and Tickle, 2011).

dorsal mesodermal genes are activated and mesodermal tissues become the organizer. *(From Agius, E., Oelgeschläger, M., Wessely, O., Kemp, C., De Robertis, E.M., 2000. Endodermal nodal-related signals and mesoderm induction in* Xenopus. *Development 127, 1173–1183.)* (B) The anteroposterior (AP) and dorsoventral (DV) axes are patterned by morphogens and their regulators in zebrafish embryos. (d) Fate map of a zebrafish gastrula with the orientation of the AP and DV axes. Neural ectoderm can be divided into four regions of the central nervous system: forebrain (FB), midbrain (MB), and hindbrain (HB), which is further subdivided into rhombomeres one to seven in zebrafish, and spinal cord (SC). (e) Wnt, fibroblast growth factor (FGF), Nodal, and retinoic acid (RA) specify posterior fates (green) in a concentration-dependent manner whereas their inhibition is required for anterior fate specification (orange). (f) The BMP morphogen gradient specifies ventral cell fates (blue) at high levels and allows dorsal fate specification (red) at low levels. BMP signaling is regulated by extracellular factors; the DV localization of their transcriptional domain is depicted. Chordin activity is key because it acts as a BMP antagonist and as the substrate for other extracellular modulators. (g) Transcriptional regulation of *bmp* and dorsal organizer genes. By activating different transcriptional repressors, zygotic Wnt promotes BMP signaling whereas FGF and Nodal antagonize it. *Blue lines* indicate activity that promotes BMP signaling and *red lines* indicate activity that limits it (*dark shades* describe a direct effect, *light shades* an indirect effect). *(From Tuazon, F.B., Mullins, M.C., 2015. Temporally coordinated signals progressively pattern the anteroposterior and dorsoventral body axes. Seminars in Cell & Developmental Biology 42, 118–133.)*

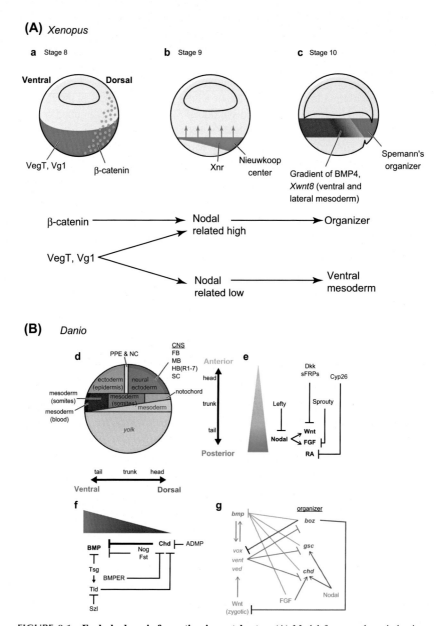

FIGURE 8.1   Early body axis formation in vertebrates. (A) Model for mesoderm induction and organizer formation in *Xenopus* embryos. (a) At the late blastula stages, Vg1 and Veg2 are found in the vegetal hemisphere; β-catenin is located in the dorsal region. (b) β-catenin acts synergistically with Vg1 and VegT to activate Nodal-related (*Xnr*) genes. This creates a gradient of Xnr proteins across the endoderm, highest in the dorsal region. (c) Mesoderm is specified by the Xnr gradient. Mesodermal regions with little or no Xnr have high levels bone morphogenetic protein (BMP)-4 and Xwnt8; they become ventral mesoderm. Those having intermediate concentrations of Xnr become lateral mesoderm. Where there is a high concentration of Xnr, *goosecoid*, and other

8.1.1 Spemann's Organizer

The dorsal marginal zone (DMZ) of amphibian embryos is a region where the involution movements of gastrulation begin (Fig. 8.1A(c)). Spemann and Mangold (1924) demonstrated that transplantation of tissue from the DMZ at the onset of gastrulation to the ventral side of a host embryo resulted in the formation of a second body axis. Ectopically induced Siamese twins contained only a few cells that came from the grafted DMZs, and most of their tissue was derived from the host embryos, indicating that the early gastrula DMZ can induce formation of a fully patterned embryo in surrounding tissue. These experiments gave rise to the term "Spemann's organizer" for the early gastrula DMZ (Fig. 8.1A(c)). Chordin, Noggin, Frzb, and other gene products act in concert as Spemann's organizer.

8.1.2 The Nieuwkoop Center

In 1967 Nieuwkoop found that vegetal cells form endoderm and animal cap explants form ectoderm when cultured in isolation, but mesoderm was not formed. Mesoderm was induced only when both types of cells were co-cultured as aggregates, suggesting that mesoderm formation requires inductive signals from vegetal cells (Nieuwkoop, 1967, 1973). In addition, Nieuwkoop also showed that there is a DV bias within the vegetal cell mass such that dorsovegetal blastomeres induce dorsal mesoderm whereas ventrovegetal blastomeres induce ventral mesoderm (Fig. 8.1A(b)). Namely, the dorsal mesoderm, including Spemann's organizer, is induced by dorsovegetal endodermal cells, and this organizer is often referred to as the "Nieuwkoop center." Nodal and fibroblast growth factor (FGF) are molecules with Nieuwkoop center activity.

8.1.3 The Three-Signal Model

These and other experiments led to the "three-signal model" of mesoderm formation: (1) vegetal cells secrete a mesendoderm-inducing factor that converts the marginal zone into a ring of mesoderm (Fig. 8.1A); (2) the vegetal cell mass is subdivided into dorsal and ventral parts that induce dorsal and ventral mesoderm, respectively (Fig. 8.1); and (3) the most dorsal mesoderm (Spemann's organizer) secretes signals that establish DV polarity by promoting dorsal identity (Fig. 8.1). The first two steps are thought to occur early in development, at the onset of zygotic transcription, whereas Spemann's organizer operates later, during gastrulation. The Nieuwkoop center contributes to the second signal in this model—it acts upstream of Spemann's organizer. Genes involved in the signals and their regulatory cascade in zebrafish embryos are shown in Fig. 8.1B (eg, Tuazon and Mullins, 2015).

8.2 AXIAL PATTERNING OF DEUTEROSTOME BODY PLANS

Details of the axial patterning of deuterostome body plans in relation to the evolution of deuterostomes and the origins of chordates were reviewed properly by Lowe et al. (2015). This section discusses this important evolutionary developmental issue according to their review.

8.2.1 The Animal-Vegetal Axis and Formation of Endomesoderm

The first event in early embryogenesis is the breaking of the egg's radial symmetry and the establishment of the An-Vg axis. This axis organizes the formation of the three germ layers: endoderm, mesoderm, and ectoderm. Ectoderm derives from the animal pole and endomesoderm from the vegetal pole. Endomesoderm later differentiates into endoderm and mesoderm. In almost all deuterostome phyla, the formation of endomesoderm is triggered by β-catenin, the intracellular effector of the canonical Wnt signaling pathway. After fertilization, β-catenin is stabilized preferentially in the vegetal pole of early embryos and activates genes of the endomesodermal cellular program. This mechanism has been demonstrated in protostomes (Henry et al., 2008) and in cnidarians (Wikramanayake et al., 2003), suggesting an early eumetazoan origin for this process. In *Xenopus*, fertilization triggers a rotation of the egg cortex relative to its cytoplasm, an event called the cortical rotation. This movement translocates β-catenin and other Wnt-related molecules to the dorsal side, and simultaneously glycogen synthase kinase-3, an intracellular antagonist of the Wnt pathway, becomes dorsally downregulated. In early *Xenopus* embryos, maternally supplied VegT and Vg1 also participate in germ layer specification (Fig. 8.1A).

Later development of mesoderm from endomesoderm basically occurs by induction in response to a signal from neighboring cells located toward the vegetal side, with the exception of ascidian muscle determinant Macho-1 (Section 10.2). In vertebrates, two main signaling pathways, Nodal and FGF, are involved in mesoderm specification (Fig. 8.1) (Agius et al., 2000; Kimelman, 2006). In amphioxus, FGF signaling specifies anterior mesoderm that forms by enterocoely (Bertrand and Escriva, 2011). In ascidians, FGF signaling is involved in notochord and mesenchyme induction (Lemaire, 2011). Likewise, in the hemichordate, *Saccoglossus kowalevskii*, FGF signaling induces mesoderm and enterocoely (Green et al., 2013). In echinoderms, Notch–Delta signaling is important in early mesoderm specification, and the role of FGF in echinoderms has yet to be fully characterized (Hinman and Davidson, 2007). Therefore, as a classic deuterostome character, it is highly likely that β-catenin is required at the start and that FGF and Nodal play a basic role in mesoderm induction, although various signals are required in later developmental processes (Fig. 8.1B).

8.2.2 The Anteroposterior Axis

Wnt signaling via β-catenin has emerged as the earliest conserved determinant of AP patterning in deuterostomes. This has been best characterized in vertebrates, in which Wnts act as posteriorizing signals in all three germ layers, although most analysis concerns central nervous system (CNS) patterning. Wnts are produced posteriorly and Wnt antagonists are produced anteriorly from anterior mesoderm. Their interaction sets up a Wnt gradient, prefiguring the eventual anatomical A/P axis. In both sea urchin larvae and *S. kowalevskii*, Wnt signaling is also important for establishing A/P patterning, suggesting that generating a Wnt signaling gradient (high posteriorly, low anteriorly) is a key step in the early development of the A/P axis in all deuterostome phyla.

The different intensities of Wnt signaling along the graded distribution then activate different toolkit genes in all three germ layers each, producing a long-lasting AP map of gene expression domains, colinear with Wnt distribution. The map reflects an enormous genetic regulatory network spread over the length of the embryo, and the domain array established by this map has been called the "invisible anatomy" of the animal (Slack et al., 1993). Relative to construction of the CNS, the AP network provides a conserved mechanism for ectoderm-derived structures among hemichordates and chordates (Section 9.2). On the other hand, in echinoderms, expression of more posterior markers such as *Hox* genes is entirely absent during early ectodermal patterning. *Hox* gene expression begins in a colinear pattern, not in ectoderm but in the posterior coelom as the adult rudiment begins to form. Comprehensive characterization of patterning of the pentaradial body plan of adult echinoderms is badly needed to unravel their evolution.

8.2.3 The Dorsoventral Dimension and the Inversion Hypothesis

The DV axis evolved in the Pre-Cambrian stem leading to the bilaterian ancestor and is intimately involved in the origin of bilateral symmetry (Figs. 8.1B, 8.2, and 8.3). One side of the embryo produces Bmp, and the opposite side produces the Bmp antagonist, Chordin (Arendt and Nübler-Jung, 1994; Holley et al., 1995; De Robertis, 2008). Recent studies of nonchordate deuterostomes suggest that Nodal regulates *Chordin* expression; therefore the BMP-Chordin gradient is sometimes called the BMP-Nodal (Chordin) gradient (Figs. 8.2 and 8.3A). Through complex interactions, this antagonism generates a graded distribution of BMP across the embryo, resulting in a graded distribution of BMP receptors and a corresponding distribution of activated Smad1/5 transcription factor in embryonic cells. The BMP gradient stimulates and represses different toolkit genes, generating a long-lasting DV map of expression domains of these genes (Fig. 8.1B); this map governs the development of different cell types and tissues along the DV axis.

There is remarkable conservation in the development of certain tissues and cell types among bilaterians. This was demonstrated by similarities in the DV development of protostomes, *Drosophila* and the annelid, *Platynereis dumerilii* (Denes et al., 2007), and vertebrates (*Xenopus*, *Mus*, and *Danio*; Holley et al., 1995; Mizutani and Bier, 2008; De Robertis, 2008). Domains from the Chordin/ Nodal side of the DV gradient activates axial (striated) muscle development in mesoderm and nerve cell development in ectoderm, especially motor neurons and interneurons that assemble into the CNS, whereas domains from the BMP side activate the heart tube and coelom development from mesoderm and endo- derm and sensory nerve cell development from ectoderm. In other words, the BMP-Chordin/Nodal gradient patterns all three germ layers.

However, a significant division appears among bilaterians that bears upon the annelid theory and the inversion hypothesis (Figs. 8.2 and 8.3). That is, the Chordin/Nodal and BMP sides of the molecular DV axis are reversed in protostomes and vertebrates. In vertebrates, the BMP side is ventral and the Chordin/Nodal side is dorsal, but in *Drosophila* and other protostomes it is the opposite (Figs. 8.1–8.3). This difference appeared to have been resolved by the inversion hypothesis, which proposed that the vertebrate ancestor underwent a DV inversion of the body relative to the substratum (Figs. 2.2 and 2.6). This transition simultaneously inverted the BMP–Chordin/Nodal axis, the domain map, and axis of anatomical differ- entiations. As a final refinement, the mouth was relocated to the BMP side, whereas most protostomes (eg, *Drosophila*) and invertebrate deuterostomes form the mouth on the Chordin/Nodal side.

A key question raised here is when the inversion occurred during bilaterian evolution. Was it at the time of the split of protostomes and deuterostomes, or did it occur at the split of ambulacrarians and chordates? Three investiga- tions of this problem using sea urchins (echinoderms), acorn worms (hemichor- dates), and lancelets (cephalochordates) have illuminated the matter. First, the BMP–Chordin/Nodal expression profile in the lancelet, *Branchiostoma flori- dae*, is similar to that of vertebrates (Fig. 8.2A) (Yu et al., 2007; Onai et al., 2010). Second, the BMP–Nodal/Chordin expression profiles in the sea urchin, *Paracentrotus lividus* (Duboc et al., 2004; Lapraz et al., 2009, 2015; Haillot et al., 2015) and the acorn worm, *S. kowalevskii* (Lowe et al., 2006) and *Ptych- odera flava* (Röttinger et al., 2015), are reversed relative to those in vertebrates and similar to those of protostomes (Fig. 8.3A). Namely, the inversion occurred during the evolution of chordates (Fig. 8.3B), and since then, the DV-axis inver- sion has been vigorously discussed as a cue to the origin of chordates (see Lowe et al., 2015).

8.2.4 Left-Right Asymmetry and the Nodal Signal

In bilaterians, LR asymmetry is essential for handed positioning, morpho- genesis, and ultimately the function of organs. It has been shown that an

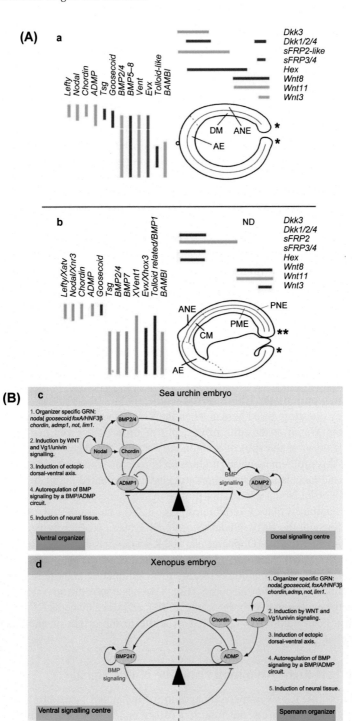

asymmetric or left-side expression of the *nodal* gene plays a pivotal role in differentiation of the LR axis of vertebrates (eg, Levin, 2005; Brennan et al., 2002). Nodal is detected originally on both sides of the node, and its expression is reinforced on the left side by Nodal flow. By a complex positive and negative feedback loop, together with *Lefty1*, *Lefty2*, *Pitx2*, and others, Nodal dictates left-sided morphogenesis of asymmetric organs, thus presaging the development of morphological asymmetries of the body (eg, Levin, 2005; Coutelis et al., 2014; Blum et al., 2014). In the context of deuterostome evolution, Nodal has been shown to act upstream of Chordin (Figs. 8.1–8.3) and as a mesodermal inducer, as discussed earlier in the case of vertebrates and as will be discussed later relative to nonchordate deuterostomes.

◀ **FIGURE 8.2 Anteroposterior (AP) and ventroposterior (VP) patterning in deutero-stomes.** (A) Schematic diagram of the expression of dorsoventral (DV) and AP patterning genes at late gastrula stage. (a) Amphioxus. (b) Frog, stage 12.5. Anterior at left; blastopore at right. Expression of DV patterning genes (irrespective of AP distribution) at left. Expression of AP patterning genes (irrespective of DV distribution) at right. *Single asterisk* indicates little involution over blastopore lip, and *double asterisk* indicates extensive involution over blastopore lip. *Green bars*, expression in both mesendoderm and ectoderm; *red bars*, expression only in mesendoderm; *blue bar*, expression in ectoderm; *gray bar*, tissue layers in which expression of AP patterning genes has not been determined. *AE*, anterior endoderm; *ANE*, anterior neuroectoderm; *CM*, chordamesoderm; *DM*, dorsal mesoderm; *ND*, not determined; *PME*, posterior mesendoderm; *PNE*, posterior neuroectoderm. *(From Yu, J.K., Satou, Y., Holland, N.D., Shin-I, T., Kohara, Y., Satoh, N., Bronner-Fraser, M., Holland, L.Z., 2007. Axial patterning in cephalochordates and the evolution of the organizer. Nature 445, 613–617.)* (B) Homologous gene regulatory networks drive the formation of the DV organizer in chordates and in the sea urchin. (c, d) Topology of the bone morphogenetic protein (BMP)-antidorsalizing morphogenetic protein (ADMP)-Chordin-Nodal network in chordates and echinoderms. (d) In chordates, BMP2/4/7 ligands are expressed in a ventral signaling center whereas *admp* is expressed on the opposite side. BMP signaling on the ventral side promotes expression of BMP ligands and represses *admp* expression (repression). Nodal signaling in the Spemann organizer promotes expression of *admp* and *chordin*. Chordin then shuttles BMP and ADMP ligands toward the ventral side where they activate BMP signaling (expansion). Chordin inhibits ADMP signaling dorsally. (c) In the sea urchin, Nodal also positively regulates the expression of chordin and of an organizer-specific BMP ligand, admp1, but Nodal also controls expression of *bmp2/4*. Chordin inhibits ADMP and BMP2/4 signaling ventrally and promotes translocation of these ligands to the opposite dorsal side where they activate BMP signaling (expansion). BMP signaling on the dorsal side does not activate *bmp2/4* expression, but induces *admp2*, which, in turn, autoregulates. BMP signaling on the side opposite from the organizer represses expression of *admp1* in the organizer by an unknown mechanism (repression). Historical continuity of the DV organizer of chordates and echinoderms is suggested by its striking conservation at the level of the whole gene regulatory network (GRN), which includes key transcription factors and signaling molecules such as *nodal, lefty, bmp2/4, chordin, admp, not, lim1*, and *HN3b/foxA*. Both GRNs require Wnt and Univin/Vg1 signaling to start, and the activity of both GRNs can endow cells with organizer-like properties, coherently inducing a whole set of tissues along the DV axis. Finally, both organizers are involved in induction of neural tissue, mainly through BMP inhibition and possibly through direct induction by Nodal in the case of sea urchin embryos. *(From Lapraz, F., Haillot, E., Lepage, T., 2015. A deuterostome origin of the Spemann organiser suggested by Nodal and ADMPs functions in Echinoderms. Nature Communications 6, 8434.)*

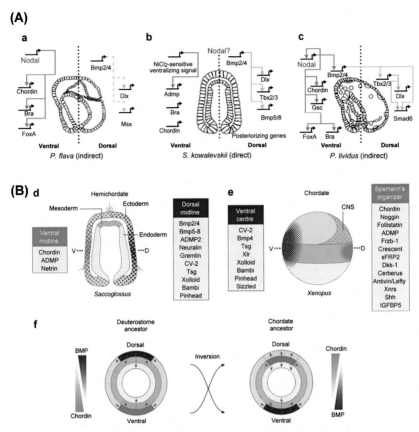

FIGURE 8.3 Genetic basis for dorsoventral (DV)-axis inversion. (A) Comparison of the role and molecular mechanisms of Nodal signaling in ambulacrarians. (a–c) Diagrams representing the molecular mechanism underlying DV patterning in ambulacrarians. (a) Hemichordate, *Pt. flava* (indirect development) and (b) *S. kowalevskii* (direct development). (c) Echinoderm, *Pa. lividus* (indirect development). *(From Röttinger, E., DuBuc, T.Q., Amiel, A.R., Martindale, M.Q., 2015. Nodal signaling is required for mesodermal and ventral but not for dorsal fates in the indirect developing hemichordate,* Ptychodera flava. *Biology Open 4, 830–842.)* (B) Comparison of DV patterning mechanisms of hemichordates and chordates. (d) BMP–Chordin signaling components expressed in dorsal and ventral midline ectoderm (blue) in the late gastrula stage of *S. kowalevskii*. (e) BMP–Chordin signaling components expressed either ventrally or dorsally in Spemann's organizer in the early gastrula of *Xenopus*. CNS, central nervous system. (f) Inversion of DV signaling centers and relocation of the Chordin source from ectoderm (yellow) to mesoderm (red) were innovations in DV patterning in ancestral chordates *(gray shading). (From Lowe, C.J., Clarke, D.N., Medeiros, D.M., Rokhsar, D.S., Gerhart, J., 2015. The deuterostome context of chordate origins. Nature 520, 456–465.)*

8.3 INTERPRETATION OF THE DORSOVENTRAL-AXIS INVERSION HYPOTHESIS

As discussed earlier, it is evident that the DV-axis inverted between nonchordate deuterostomes and chordates. BMP-Nodal expression is opposite between them, BMP being expressed on the dorsal (oral) side in nonchordate deuterostomes and on the ventral (aboral) side in chordates. At the risk of being repetitive, the inversion hypothesis alone cannot explain chordate novelties. On the other hand, the new organizers hypothesis thus far cannot explain how the DV-axis inversion occurred. Before further discussion, the role of Nodal in mesoderm formation should be mentioned.

8.3.1 Nodal Function in Mesoderm Formation

In vertebrates, Nodal functions in endomesoderm formation (Fig. 8.1). In sea urchins, *nodal* is expressed exclusively along the venter throughout early embryogenesis (Figs. 8.2B(c) and Fig. 8.3A(c)) (Duboc et al., 2004; Smith et al., 2008). Knockdown of *nodal* function perturbs the establishment of the DV and LR axes. Although endoderm and mesoderm form, patterning of the two germ layers is affected. Interestingly, *Brachyury* expression around the mouth-opening diminishes in nodal-deficient embryos. In addition, a recent study shows that Nodal and antidorsalizing morphogenetic protein (ADMP; an atypical BMP ligand) coordinarily act to induce the secondary axis in the same manner as Spemann's organizer (Lapraz et al., 2015). In hemichordates, a recent study of *P. flava* shows that during gastrulation, *nodal* is expressed in a ring of cells at the vegetal pole that gives rise to endomesoderm and in the ventral ectoderm at later stages of development (Fig. 8.3A(a)) (Röttinger et al., 2015). Inhibition of Nodal function disrupts DV fates and blocks formation of only mesodermal, apical, and ventral gene expression, but not dorsal gene expression. This study suggests that cooption of Nodal signaling in mesoderm formation occurred before the emergence of chordates and that Nodal signaling on the ventral side is uncoupled from BMP signaling on the dorsal side (Fig. 8.3A). In summary, Nodal plays pivotal roles in mesoderm formation and DV-axis formation in nonchordate deuterostomes and chordates.

8.3.2 Interpretation of the Dorsoventral-Axis Inversion

As pointed out by Lowe et al. (2015), the Nodal and BMP sides of the molecular DV axis have different anatomical names in nonchordate deuterostomes and vertebrates. In vertebrates, the BMP side is called "ventral" and the Nodal side is "dorsal," but in ambulacrarians the molecular and anatomical links are reversed. By zoological convention, sides are named according to the animal's orientation to the substratum and the location of the mouth of the adult.

As a result of discussions with Yuuri Yasuoka, Jr-Kay Yu, and others, I present an alternative interpretation of how the mouth opening was reversed during the origin of chordates. It is possible to consider the inversion hypothesis from a different point of view—not by asking how the DV axis was inverted but by asking where and how the mouth opening position was determined in nonchordate and chordate embryos.

As shown in Fig. 8.4, at the blastula and early gastrula stages of the deuterostome stem and extant echinoderms and hemichordates, the mouth-opening position (MOP) in endoderm is located at the vegetal pole (white asterisks in Fig. 8.4A and B). At the late gastrula stage, the tip of the archenteron vents to the side of ectoderm where Nodal is expressed (Fig. 8.4C). Therefore MOP corresponds to the Nodal site. Accordingly, the Nodal site is ventral. On the other hand, at the blastula stage of the chordate stem, the An-Vg axis was distorted approximately 20–30°, leading to an apparent inclination of that

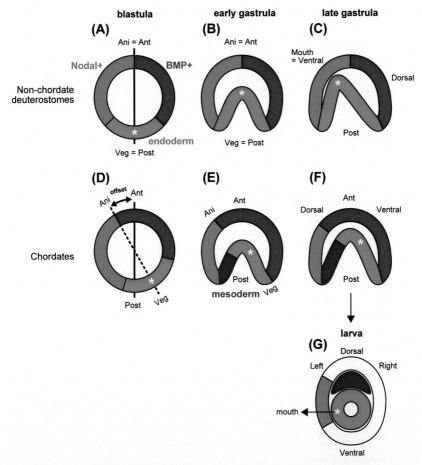

FIGURE 8.4   Schematic representation for interpretation of the DV-axis inversion between nonchordate deuterostomes and chordates. See text for details. Embryonic regions marked by blue, red, green, and dark blue are Nodal+ region, bone morphogenetic protein (BMP)+ region, endoderm, and mesoderm region, respectively. *Ani*, animal pole; *Ant*, anterior pole; *Post*, posterior pole; *Veg*, vegetal pole. *White asterisks* indicate mouth-formation sites. *Original drawing of Yuuri Yasuoka.*

axis compared with the original AP axis (Fig. 8.4D). This was demonstrated by a comparison of fate map development in amphioxus and vertebrates (Holland and Holland, 2007) and by slightly vegetal-side-forwarded expression of *Nodal* in amphioxus late blastulas (Onai et al., 2010). This inclination was also suggested by a recent study of asymmetric localization of presumptive primordial germ cells expressing *Vasa* and *Nanos* (Wu et al., 2011). Because of this An-Vg axis distortion, MOP appears to incline toward the original BMP side (Fig. 8.4D). At the next step (gastrulation), presumptive mesodermal cells with Nodal expression are invaginated together with the presumptive endodermal cells (Fig. 8.4E and F). Mesendoderm formation begins at the Nodal side, leaving less space for MOP. As a consequence, the mouth might have to open on the BMP side; therefore the BMP side becomes ventral (Fig. 8.4G). Of course, further studies are needed to evaluate this possible explanation, but it may be that in reality no DV inversion occurs in nonchordate deuterostomes and vertebrates.

Distortion of the An-Vg axis in early embryos must be a significant developmental event for chordate origins and evolution because amphibians and ascidian urochordates developed a more sophisticated system for the distortion. In *Xenopus*, the sperm entry point becomes a center for cortical contraction of fertilized eggs, by virtue of cytoskeletal components, mainly microtubules (eg, Weaver and Kimelman, 2004). Cortical movement results in formation of the gray crescent, which in turn becomes the dorsal side of the embryo. Relying on these events, the radial symmetry is converted to early bilateral symmetry; one side is dorsal and the other side is ventral. This asymmetry causes an unequal stabilization of maternally provided β-catenin, higher at the dorsum and lower at the venter (Fig. 8.1A), and then initiates the DV gradient supported by BMP-Chordin antagonistic expression and function. Similarly, in ascidian eggs, sperm entrance evokes cortical rotation to form the myoplasm, which marks the future dorsal side of the embryo (see Satoh, 2014).

8.4 CONCLUSIONS

Among many molecular components involved in body axis formation, Nodal signaling is likely to have special functions because *nodal* plays multiple roles in the deuterostome body plan, first for establishment of left-right asymmetry (its original function), second for DV patterning, and third for early mesoderm formation. As mentioned earlier, if the DV axis was inverted, did the echinoderm and/or hemichordate stems evolve to chordates? The answer is evident. They could not have led to chordates. We need to seek a better interpretation, such as the new organizers hypothesis.

Chapter 9

The Enteropneust Hypothesis and Its Interpretation

Among the hypotheses of chordate origins discussed heretofore, the entero-pneust hypothesis is unique in its emphasis on the similarities of adult anatomy and organogenesis between hemichordates and chordates. These similarities have received support from expression profiles of genetic toolkit genes. The new organizers hypothesis also focuses on enteropneusts hemichordates and chordates derived from the common ancestor of deuterostomes but having evolved independently as discrete lineages. How do we interpret this similarity to better understand chordate origins?

In previous chapters, in light of recent advances in molecular phylogeny, comparative genomics, and evolutionary developmental biology, we reexamined hypotheses proposed to explain chordate origins. As a result, most of these have now lost credibility because of the way they interpret deuterostome evolutionary scenarios. One remaining hypothesis for the chordate origins, the enteropneust hypothesis, is unique in that it emphasizes the similarities of hemichordate adult anatomy and organogenesis to those of chordates (Cameron et al., 2000; Gerhart, 2000, 2001; Holland, 2000; Gerhart et al., 2005; Gerhart, 2006; Brown et al., 2008; Satoh, 2008). Alexander Kowalevsky (1866a) identified chordate-like pharyngeal gill slits in acorn worms. William Bateson (1886) suggested an evolutionary link of the stomochord to the notochord, and Thomas H. Morgan (1894) discussed the similarity of collar cord development and chordate neurulation. Organogenesis of these structures has been the target of further anatomical discussions (eg, Ruppert, 2005) and intensive studies of the expression profiles of genetic toolkit genes that are expressed in related organs and tissues of chordates (eg, Lowe et al., 2003, 2006; Pani et al., 2012; Miyamoto and Wada, 2013; reviewed by Lowe et al., 2015).

The new organizers hypothesis assumes that hemichordates and chordates were both derived from the common ancestor of deuterostomes, but that they evolved independently, each leading to a discrete lineage. The anatomical similarity so far discussed includes the pharyngeal gill slits, the stomochord, and the nervous system. Are these homologies or homoplasies (convergent evolution)? Because the pharyngeal gill slit is a synapomorphy of deuterostomes or homologous between hemichordates and chordates, this chapter will discuss the stomochord and the nervous system of hemichordates in relation to chordate origins.

Chordate Origins and Evolution. http://dx.doi.org/10.1016/B978-0-12-802996-1.00009-2

9.1 THE STOMOCHORD AND OTHER ORGANS PROPOSED AS ANTECEDENTS TO THE NOTOCHORD

9.1.1 The Stomochord

As discussed in Section 2.6 of Chapter 2, the stomochord has attracted much interest as a notochord-like ancestral trait, ever since Bateson (1886) demonstrated it as a "hemi-chord." It is a diverticulum of the anterior gut that extends into the posterior proboscis and is thought to support the heart–kidney complex along its dorsal surface (Fig. 9.1B; Fig. 1.3B). Stomochord cells are vacuolated and surrounded by a sheath, similar in tissue organization to a notochord (Balser and Ruppert, 1990). However, the homology of the stomochord and notochord is less well supported by morphological and molecular criteria (Satoh et al., 2014b).

First, as shown in Table 9.1, the developmental mode of the stomochord differs from that of the notochord. The former is formed during the juvenile stage by extension of the anterior gut whereas the latter is formed at the gastrula and neurula stages by convergent/extension movements of precursor cells. Second, *Brachyury*, a key regulatory gene for notochord formation (Section 7.5.1 of Chapter 7), is not expressed in the stomochord-forming region (Peterson et al., 1999a). Third, on the other hand, *FoxE* is expressed rather specifically in the stomochord-forming region in the juvenile acorn worm. The *FoxE* gene is expressed in the club-shaped gland and in the endostyle of amphioxus, in the endostyle of ascidians, and in the thyroid gland of vertebrates (Satoh et al., 2014b). Fourth, the anterior endodermal location of the stomochord corresponds to that of pharynx-related organs of chordates. On the basis of these findings, it is highly likely that the stomochord has an evolutionary affinity to chordate organs derived from the anterior pharynx, but not to the notochord (Table 9.1). That is, the stomochord/notochord relationship is one of superficial similarity or homoplasy between hemichordates and chordates.

9.1.2 The Pygochord

The pygochord is a stiff vacuolated rod in the ventral midline of the posterior trunk of ptychoderid hemichordates and runs between the ventral blood vessel and the intestine. In the context of the inverted enteropneust hypothesis, the pygochord was suggested to be homologous to the notochord (Nübler-Jung and Arendt, 1996; Annona et al., 2015). Further investigations of the pygochord are needed because we have almost no molecular developmental data on this organ to date. However, as in the case of the stomochord, the embryonic stage in which the two organs form is completely different. The notochord is formed during early embryogenesis whereas the pygochord arises during adult organogenesis. In addition, the enteropneust hypothesis does not pay much attention to axis inversion because the stomochord and the major enteropneust nervous system exist dorsally in the adult; it does not require dorsoventral (DV)-axis inversion.

(A)

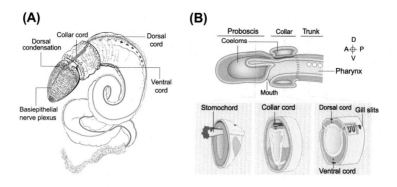

(B)

(C)

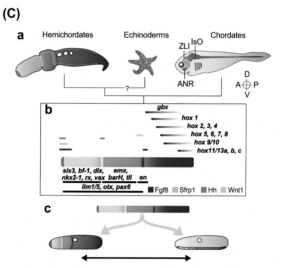

FIGURE 9.1 **Features of enteropneust and chordate affinities.** (A, B) Key anatomical features of the enteropneust body plan. (A) The nervous system of an adult enteropneust showing the broad basiepithelial plexus throughout the ectoderm and nerve chords along the dorsal and ventral midlines. *Blue spots* represent cell bodies and lines represent neural processes. (B) Longitudinal and transverse sections through an adult enteropneust hemichordate, highlighting morphological characters that have featured prominently in classic hypotheses of deuterostome evolution and chordate origins. *A,* anterior; *D,* dorsal; *P,* posterior; *V,* ventral. (C) A conserved molecular network for the deuterostome anteroposterior (AP) axis. (a) Schematic representation of the distribution of ectodermal expression domains of AP transcription factors (*blue gradient*) and ectodermal signaling centers (*green, yellow,* and *red*) in relation to the body plans of deuterostome phyla. Chordate neuroectodermal signaling centers depicted are the anterior neural ridge (ANR), zona limitans intrathalamica (ZLI), and isthmic organizer (IsO). Broad conservation of expression domains between hemichordates and chordates allows for the reconstruction of an ancestral patterning network, which is shown without any explicit inference of ancestral morphologies. Insufficient data exist from echinoderms to infer to what extent they share this conserved AP patterning network during adult patterning, although much of the earlier network is conserved in larvae. (b) Domain map for the conserved transcription factors and signaling ligands in relation to the AP axis. (c) Current data allow for the reconstruction of a conserved molecular coordinate system for AP axis of the last common deuterostome ancestor, but not for the reconstruction of discrete morphologies of that ancestor, because this AP patterning network is deployed in various morphological contexts, as evidenced by comparative data from hemichordates (dispersed; AP expression domains encircling the body) and chordates (condensed; AP domains largely restricted to regions near the dorsal midline). *A,* anterior; *D,* dorsal; *P,* posterior; *V,* ventral. *From Lowe, C.J., Clarke, D.N., Medeiros, D.M., Rokhsar, D.S., Gerhart, J., 2015. The deuterostome context of chordate origins. Nature 520, 456–465.*

TABLE 9.1 Comparison of Characteristic Features Between the Acorn Worm Stomochord and Chordate Notochord

Organ	Stomochord	Notochord	
		Cephalochordates	Urochordates
Developmental stage at which the organ is formed	Juveniles	Embryos	Embryos
The region in which the organ is formed	Proboscis of juveniles (anterior region of the body)	Dorsal midline of fish-like larvae (anterior to posterior region of the body)	Tail of tadpole-like larvae (posterior region of the body)
Developmental and morphological features	Formed by anterior outgrowth of pharynx (buccal diverticulum); layered cells, vacuolated; covered with extracellular matrix	Formed by pinching off of dorsal archenteron; coin-shaped cells with myofibrils; covered with the notochord sheath	Formed by convergence, intercalation, and extension of precursor cells; vacuolated; covered with the notochord sheath
Gene Expression Profile			
Brachyury	−	+	+
FoxE	+	−	−

9.1.3 The Annelid Axochord

In relation to the annelid and inversion hypotheses, Lauri et al. (2014) described the axochord in the polychaete annelid, *Platynereis dumerilii*, as a possible notochord precursor. The axochord is a midline mesodermal structure of muscle cells contained in a strong sheath, to which lateral muscles attach. This study reports a suite of genes with conserved roles in notochord development, including *Brachyury*, which is expressed in the annelid's developing ventral mesoderm and ectoderm. Therefore the axochord does have some structural and functional similarities to the chordate notochord (Medeiros, 2015; Annona et al., 2015). However, as discussed in Section 7.5 of Chapter 7, the notochord is an organ for fish-like larval development of chordates. The axochord may facilitate the creeping locomotion of annelid worms; nevertheless, development of the axochord in annelids does not appear to be directly related to emergence of the chordate notochord (Hejnol and Lowe, 2014). Discussion of this relationship may be meaningful in relation to bilaterians in general, but it seems irrelevant to the origins of chordates.

9.2 THE NERVOUS SYSTEM OF ENTEROPNEUSTS

The nervous system of adult hemichordates is more complex than the nervous system of adult echinoderms and more closely resembles the chordate central nervous system (CNS; see recent review of Lowe et al., 2015 and references therein). As shown in Fig. 9.1A, the nervous system of acorn worm adults displays two organizational features: a broad basiepithelial plexus, particularly prominent in proboscis ectoderm, and a pair of nerve cords (ventral and dorsal). The ventral cord extends the length of the trunk and the dorsal cord runs from the base of the proboscis down the length of the animal and connects to the ventral cord via lateral nerve rings (Fig. 9.1A). Both cords are superficial condensations of the nerve plexus except in a short length that spans the collar (Ruppert, 2005). In the collar region, the cord is internalized into a tube with a prominent lumen in some species (Fig. 9.1B). It is formed by ingression of dorsal epidermal cells, a developmental process that resembles chordate neurulation (Morgan, 1891, 1894; Kaul and Stach, 2010; Luttrell et al., 2012; Miyamoto and Wada, 2013).

9.2.1 The Dorsal Collar Cord

Various authors have proposed ventral and dorsal collar cords as possible homologous structures of the chordate dorsal nerve cord, with the internalized collar cord having attracted the most attention. Although early reports suggested that the dorsal cord was simply a conduit for axons, recent studies indicate that condensations of cell bodies are associated with this cord (eg, Nomaksteinsky et al., 2009). A further molecular study in *Balanoglossus simodensis* revealed that *bmp2/4* and *pax3/7* are expressed in the collar cord (Miyamoto and Wada, 2013). The gene expression profile is similar to that of the most lateral edges of the vertebrate neural plate during CNS development. On the other hand, in *Saccoglossus kowalevskii*, markers of medial rather than distal neural plate are not expressed in the dorsal cord as predicted but rather along the ventral midline, associated with the ventral cord (Lowe et al., 2006). In addition, in *Saccoglossus*, several neural markers are not only expressed in the collar cord but also throughout the length of the superficial cord in the trunk, suggesting a patterning role of these genes along the dorsal midline (Lowe et al., 2006).

Such differences among enteropneusts in gene expression profiles related to the general organization of the nervous system forced Lowe et al. (2015) to state that no simple homology statements can yet be made about a homolog of hemichordate nervous systems in relation to those of chordates. They further argue that, although it seems likely that ancestral deuterostomes inherited some elements of nervous system centralization from the bilaterian common ancestor, a comprehensive characterization of key molecular markers is needed to further test competing hypotheses of nervous system evolution (Lowe et al., 2015). It remains unclear whether the main features of the unusual enteropneust nervous system can be ascribed to the filter-feeding deuterostome ancestor, modified thereafter in the chordate lineage, or whether they are secondary derivatives of the hemichordate lineage.

9.2.2 Gene Expression Profiles in Epidermis along the Anteroposterior Axis

As discussed in Section 8.2.2 of Chapter 8, a Wnt signaling gradient (high posteriorly, low anteriorly) is a key step in anteroposterior (AP)-axis formation in chordates for adult and larval body plans. The AP network provides an especially conserved mechanism for ectoderm-derived structures among hemichordates and chordates in the construction of the CNS (see Fig. 7.7). In particular, striking similarities of genetic toolkit gene expression along the AP axis have been discovered in *S. kowalevskii* (Fig. 9.1C; Lowe et al., 2003; Pani et al., 2012). That is, in the most anterior regions, coexpression of genes such as *sfrp1/5*, *fgf8/17/18*, *foxG*, retinal homeobox, *dlx*, and *nk2-1* define ectodermal territories that later form the proboscis in hemichordates and the forebrain in vertebrates. Further to the posterior, expression domains of *emx*, *barH*, *dmbx*, and *pax6* define the collar of hemichordates and the midbrain of vertebrates. More posteriorly still, domains of *gbx*, *engrailed*, *pax2/5/8*, and the colinearly expressed *Hox* genes develop the hindbrain and spinal cord in vertebrates and the pharynx and trunk of hemichordates (Fig. 9.1C). As discussed in Section 6.3.1 of Chapter 6, the *Hox* genes of hemichordates exist as an intact cluster, as in chordates (Fig. 6.3A).

9.2.3 Gene Expression Profiles along the DV Axis

As discussed in Section 8.2.3 of Chapter 8, the DV axis is established by opposing Nodal and bone morphogenetic protein (BMP) signaling gradients, with Chordin-mediated BMP inhibition in the dorsal ectoderm, segregating the presumptive CNS from epidermal (or general) ectoderm (see Fig. 7.6). Along the DV axis of the CNS, all chordates have molecularly distinct domains, including a dorsal domain, a ventral floor plate expressing hedgehog ligands, and an intervening bilateral domain flanking the neural tube lumen. Although *pax3/7* and *pax2/5/8* are expressed in different regions along the DV axis of the collar cord (Miyamoto and Wada, 2013), the DV-axis–related gene expression profiles of hemichordate nervous system require further investigation.

9.3 THE SPEMANN'S ORGANIZER-LIKE SYSTEM IN HEMICHORDATES

In vertebrates, the embryonic region of developing notochord is a key source of the secreted BMP antagonists, Chordin, Noggin, and Follistatin, and the ventralizing ligand, Shh. Using *B. simodensis*, Miyamoto and Wada (2013) examined development of the collar nerve cord in relation to endoderm development beneath the cord. They found similar expression patterns of CNS-related genes during neurulation in hemichordates and chordates. The dorsal endoderm of the buccal tube and the stomochord especially express *hedgehog*, and the collar cord expresses its receptor, *patched*, suggesting that overlying collar cord cells can receive the signal. They suggest that this endoderm likely functions as Spemann's organizer to pattern the overlying collar cord, similar to the relationship between the notochord and the neural tube in chordates. Thus it has been argued that the origin of core genetic mechanisms for development of the notochord and the neural tube date back to the last common deuterostome ancestor.

9.4 INTERPRETATIONS OF THE ENTEROPNEUST HYPOTHESIS

As described earlier, recent studies of genetic toolkit gene expression profiles have illuminated an evolutionary link in organ formation between hemichordates and chordates. However, more careful examination and interpretation of the results will likely result in a better interpretation thereof. The most striking resemblance is seen in expression profiles of chordate CNS-related genes along the AP axis of adult hemichordates (Fig. 9.1C). In addition, the enteropneust expression profile of genes involved in the vertebrate midbrain–hindbrain boundary organizer also closely resembles that of vertebrates (Fig. 9.1C). Namely, hemichordates and chordates appear to utilize very similar toolkit genes for regional patterning of the nervous system along the AP axis. However, in hemichordates all of these genes are expressed in ectoderm surrounding the entire body, indicating that the hemichordate pattern is dispersed, in contrast to the concentrated pattern of chordates (Fig. 9.1C). This raises many questions. For instance, why do enteropneusts not construct a centralized nervous system as chordates do because both are able to express CNS-related genes with very similar patterns? Temporal differences of CNS-related gene expression also require attention. In previous studies, hemichordate CNS-related gene expression was demonstrated during the juvenile formation stage (eg, Lowe et al., 2003) whereas chordate genes are expressed during the neurula stage. A recent study demonstrated that a set of neurogenesis-related genes is expressed at the blastula and gastrula stages of *S. kowalevskii* (Cunningham and Casey, 2014). Again, one of the major questions is that hemichordates and chordates express a very similar suite of toolkit genes but with very different results. Together with other data, these findings may indicate that hemichordate organogenesis is not as similar to that of chordates as was thought when their genetic toolkit gene expression profiles were first discovered.

There are two ways to interpret the enteropneust hypothesis. One is more general and is adopted by most deuterostome evolutionary developmental biologists (eg, Miyamoto and Wada, 2013). Namely, during deuterostome evolution, the enteropneusts developed various organs that are homologous to chordate homologs; therefore comparison of these organs provides clues for consideration of chordate origins.

In contrast, according to other interpretations, such as the new organizers hypothesis, these similarities appear to be superficial or to have resulted from convergent evolution. It is likely that bilaterians evolved or diverged from the common ancestor(s) during a very limited period of time (not gradually over the last 500 million years). In addition, a strong evolutionary constraint affected their body plan evolution because all metazoans existed under the same conditions (eg, Erwin et al., 2011). Therefore their basic structure along the AP, DV, and left-right axes is similar. No extant bilaterians exhibit structures that conflict with basic principles of body plan development. If such deviants arose during trials of body plan modification, then they apparently disappeared without leaving fossils. A very similar suite of genetic toolkits was used for construction of the basic body, but it resulted in very different bodies in different lineages.

9.5 CONCLUSIONS

The most striking discovery of evolutionary developmental biology is that various metazoan morphologies are constructed using a similar set of transcription factors and signal pathway molecules. The similarity of expression profiles of genetic toolkit genes in organs of metazoans belonging to different taxa reveals the homology of those organs and attests to the evolutionary relatedness of different metazoans. On the other hand, it is also common that specific suites of genetic toolkit genes have been co-opted in the formation of different organs and tissues in discrete lineages of metazoans. It is now time to ask why metazoans are so diverse despite using similar toolkit genes for body construction.

Chapter 10

Chordate Evolution: An Extension of the New Organizers Hypothesis

After their appearance, chordates evolved into three independent lineages: cephalochordates, urochordates, and vertebrates. Cephalochordates probably have retained their original form for more than 500 million years. On the other hand, urochordates and vertebrates developed unique morphologies to adapt as advanced filter feeders with tunics and specialized predators with heads and jaws, respectively. The fossil record suggests that vertebrate evolution was protracted.

As discussed in previous chapters, chordates originated from a common ancestor shared by all deuterostome taxa, approximately 570 million years ago (MYA). Among various hypotheses to explain chordate origins, I propose the new organizers hypothesis, emphasizing that the occurrence of fish-like larvae, which required a dynamic modification of larval form, is a critical evolutionary event. Current molecular data suggest that the origin of chordates should not be seen as a consequence of ambulacrarian evolution, especially that of enteropneust hemichordates. Although similar sets of toolkit genes have been co-opted in several developmental processes critical to each lineage, the evolution of hemichordates and chordates should be considered as completely independent evolutionary events.

Once the chordate ancestor (most likely similar to an extant lancelet) appeared, it apparently diverged to give rise to three different taxa: cephalochordates, urochordates (tunicates), and vertebrates. From a morphological point of view, cephalochordates have likely retained their original form for more than 500 million years (MY), presumably because of a good match between body plan and lifestyle. On the other hand, urochordates evolved a uniquely different body plan with a tunic and adapted to become advanced filter feeders, whereas vertebrates developed heads and jaws and adopted predatory lifestyles (including herbivores). Recent evolutionary developmental biology studies using *Ciona intestinalis* embryos and sophisticated techniques suggest genetic mechanisms involved in tunicate and vertebrate evolution. No paleontological data are available to infer how long tunicate evolution required, but the fossil record suggests that vertebrate evolution required a longer period. Chordate evolutionary events,

Chordate Origins and Evolution. http://dx.doi.org/10.1016/B978-0-12-802996-1.00010-9
143

especially those of vertebrates, are numerous enough to require an additional book. Therefore I will discuss them only briefly here.

10.1 EVOLUTION OF VERTEBRATES

Conservation of anatomy between cephalochordates and vertebrates (Chapter 1), genomic synteny (Chapter 5), and gene toolkit expression profiles (Chapter 7) strongly suggest that vertebrates evolved from a lancelet-like ancestor. The nomenclatural convention designating cephalochordates as "acraniates" and vertebrates as "craniates" highlights the fact that vertebrates evolved by developing a head and jaws, which enabled the transition from filter feeding to active predation (eg, Shimeld and Holland, 2000; Holland and Holland, 2001; Shigetani et al., 2005). Gans and Norticutt (1983) proposed that evolutionary development of the vertebrate head was dependent on the neural crest and neurogenic placodes that contribute to the development of the cranium and associated sensory organs. In addition, vertebrate-specific innovation includes the midbrain–hindbrain boundary organizer (Section 7.8.2), the heart, cartilage, and the adaptive immune system. These will be described briefly in the following subsections.

10.1.1 Neural Crest

The neural crest comprises a remarkable cell type, both for their ability to migrate extensively and for their capacity to differentiate into numerous derivatives (see reviews of Le Douarin and Kalcheim, 1999; Yu, 2010; Simões-Costa and Bronner, 2015; Green et al., 2015; related references therein). Because of these features, the neural crest is sometimes considered a "fourth germ layer" (Hall, 2000). Neural crest cells are induced in the ectodermal germ layer during gastrulation and initially reside at the neural plate border, positioned at the lateral edges of the central nervous system (CNS; Fig. 10.1A(a–b)). During neurulation, this border elevates as the neural plate closes to form the neural

FIGURE 10.1 **Neural crest formation in chordates.** (A) The neural crest is a multipotent cell population. (a) Schematic dorsal view of a 10-somite-stage chicken embryo, showing the neural crest (*green*) in the vicinity of the midline. *Dotted lines* delimit the embryonic region represented in cross-section (b–e). (b) Development of the neural crest begins at the gastrula stage, with the specification of the neural plate border at the edges of the neural plate. (c) As the neural plate closes to form the neural tube, neural crest progenitors are specified in the dorsal part of the neural folds. (d) After specification, neural crest cells undergo an epithelial-to-mesenchymal transition and delaminate from the neural tube. (e) Migratory neural crest cells follow stereotypical pathways to diverse destinations, where they give rise to distinct derivatives. (f) The neural crest has the capacity to give rise to diverse cell types, including mesenchymal, neuronal, secretory, and pigmented cells. *(From Simões-Costa, M., Bronner, M.E., 2015. Establishing neural crest identity: a gene regulatory recipe. Development 142, 242–257.)* (B, C) Gene regulatory interactions controlling vertebrate neural crest formation and the tunicate a9.49 cell lineage. (B) A neural crest gene regulatory network (GRN) endows this cell population with its unique features. This GRN is composed of different modules arranged hierarchically, which control each step of neural crest development (C). Regulatory circuit of a tunicate neural-crest (NC)-like pigmented cell precursor. *(From Green, S.A., Simões-Costa, M., Bronner, M.E., 2015. Evolution of vertebrates as viewed from the crest. Nature 520, 474–482.)*

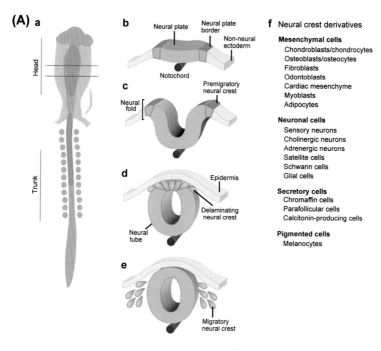

(A) a

Head

Trunk

b

Neural plate

Neural plate border

Non-neural ectoderm

Notochord

c

Neural fold

Premigratory neural crest

d

Epidermis

Delaminating neural crest

Neural tube

e

Migratory neural crest

f Neural crest derivatives

Mesenchymal cells
Chondroblasts/chondrocytes
Osteoblasts/osteocytes
Fibroblasts
Odontoblasts
Cardiac mesenchyme
Myoblasts
Adipocytes

Neuronal cells
Sensory neurons
Cholinergic neurons
Adrenergic neurons
Satellite cells
Schwann cells
Glial cells

Secretory cells
Chromaffin cells
Parafollicular cells
Calcitonin-producing cells

Pigmented cells
Melanocytes

(B) Vertebrate neural crest GRN

Signalling module

Neural-plate-border module

Neural-crest-specification module

Neural-crest-migration module

Differentiation gene batteries

WNTs   BMPs

Notch   FGFs

Zic   Msx   Gbx2

Pax3/7   Dlx5/6   Ap2

FoxD3   Snai   Ap2   SoxE

ID   Pax3/7   Ets1

FoxD3   SoxD   SoxE

RxrG   Lmo4   Ap2

Phox2b   Sox9   Mitf

Neurons and glia   Chondroblasts and osteoblasts   Melanocytes

(C) Tunicate NC-like cell circuit

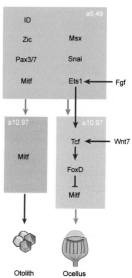

ID   a9.49

Zic   Msx

Pax3/7   Snai

Mitf   Ets1 ◄── Fgf

a10.97   a10.97

Mitf

Tcf ◄── Wnt7

FoxD

Mitf

Otolith   Ocellus

tube (Fig. 10.1A(c–d)). Nascent neural crest cells develop in the dorsal aspect of the juxtaposed neural folds. After neural tube closure/cavitation, these cells subsequently leave the CNS via an epithelial-to-mesenchymal transition (Fig. 10.1A(d)), resulting in their transformation into a multipotent, migratory progenitor cell population (Fig. 10.1A(e)) that undergoes some of the longest migrations of any embryonic cell type, often moving along highly diversified pathways.

After migration, neural crest cells progressively differentiate into distinct cell types according to environmental influences encountered during their journey and at their destinations, where they cooperate with other cell populations to form appropriate tissues and organs (Fig. 10.1A(f); Bronner and LeDouarin, 2012; Steventon et al., 2014). Cell lineage-tracing experiments have identified the derivatives and migration pathways of various neural crest populations (Le Douarin, 1982). Regionalization of the body axis from anterior to posterior is reflected by neural crest subpopulations that contribute to some overlapping as well as axial level-specific derivatives. For example, cranial neural crest cells form the facial skeleton, including the upper and lower jaw and bones of the neck as well as the glia and some neurons of the cranial sensory ganglia (Fig. 10.1A(f)). Just below the head, vagal neural crest cells populate the outflow tract of the heart and enteric ganglia of the gut. Trunk neural crest contributes to the dorsal root and sympathetic ganglia of the peripheral nervous system (Fig. 10.1A(f)), migrating in a segmental pattern through the somites. However, the trunk crest normally lacks the ability to form cartilage or bone or to contribute to the cardiovascular system. Pigment cells of the skin and peripheral glia arise from neural crest cells at all axial levels (Fig. 10.1A(f)).

10.1.1.1 Evolution of the Neural Crest and Its Gene Regulatory Network

The gene regulatory network (GRN) underlying neural crest formation appears to be highly conserved as a vertebrate innovation (Fig. 10.1B; Meulemans and Bronner-Fraser, 2004; Simões-Costa and Bronner, 2015; Green et al., 2015). Border induction signals [bone morphogenetic protein (BMP) and Fgf] from ventral ectoderm and underlying mesendoderm pattern the dorsal ectoderm, inducing expression of neural border specifiers (*Zic* and *Dlx*). These inductive signals then work with neural border specifiers to upregulate expression of neural crest specifiers (*SoxE*, *Snail*, and *Twist*). Neural crest specifiers cross-regulate and activate various effector genes (*RhoB* and *Cadherins*), each of which mediates a different aspect of the neural crest phenotype, including cartilage (*Col2a*), pigment cells (*Mitf*), and peripheral neurons (*cRet*; Fig. 10.1B).

The cephalochordate, amphioxus, lacks a neural crest. It has been shown that amphioxus lacks genes for neural specifiers and the effector subcircuit controlling neural crest delamination and migration (Fig. 10.1B; Yu, 2010; Simões-Costa and Bronner, 2015). On the other hand, the presence of a neural crest in ascidian urochordates has been debated (Jeffery et al., 2004). A recent study of

Ciona embryos demonstrates that the neural crest melanocyte regulatory network predated the divergence of tunicates and vertebrates, but the co-option of mesenchyme determinants, such as *Twist*, into neural plate ectoderm is absent (Fig. 10.1C; Abitua et al., 2012). Therefore it may be said that the neural crest evolved as a vertebrate-specific GRN innovation. This is probably associated with vertebrate development of a head and jaws.

10.1.2 The Placodes

Neurogenic placodes contribute to cranial sensory systems that mediate hearing, smell, and taste in vertebrates (see reviews of Shimeld and Holland, 2000; Patthey et al., 2014; Schlosser et al., 2014; Diogo et al., 2015; related references therein). They are considered a vertebrate innovation because the full repertoire, including olfactory, otic, epibranchial, and trigeminal placodes, are also present in jawless hagfish and lampreys. As in the case of neural crest, the development of neurogenic placodes is likely to have been a crucial event to facilitate the transition from filter-feeding invertebrate chordates to predatory vertebrates.

Eight types of placode include the adenohypophyseal placode, olfactory placodes, lens placodes, trigeminal placodes, profunda placodes, lateral line placodes, otic placodes, and epibranchial placodes (Fig. 10.2). The adenohypophyseal placode gives rise to the anterior pituitary, the major hormonal control organ of vertebrates, with six types of endocrine cells: gonadotropes, thyrotropes, corticotropes, melanotropes, lactotropes, and somatotropes. The olfactory placode generates chemosensory neurons (primary sensory cells with an axon) of the olfactory epithelium and the vomeronasal organ. Olfactory sensory neurons form a heterogeneous population. The best known are the neurons producing gonadotropin-releasing hormone (GnRH), which controls secretion of gonadotropins, and neuropeptide Y neurons, which in turn, regulate GnRH secretion. The lens placode develops into cells that become translucent by accumulation of crystallins and form the lenses of the eyes. The profundal and trigeminal placodes generate somatosensory neurons mediating temperature, touch, and pain sensation in the head. Lateral line placodes generate very similar hair cells and afferent neurons of the lateral line system used to detect water movements in many aquatic vertebrates. The otic placode generates mechanosensory hair cells (secondary sensory cells without axons) and their afferent neurons. Epibranchial placodes generate viscerosensory neurons, which, as their name indicates, innervate sensory organs associated with the digestive tract and its derivatives (Fig. 10.2).

10.1.2.1 The Evolution of the Placode and Its Gene Regulatory Network

As discussed in Section 7.9, for the last 150 years, debate has swirled around the question of whether placodes are a vertebrate innovation or whether there are placode homologs in cephalochordates and urochordates. The relationship

(A)

(B)

(C)

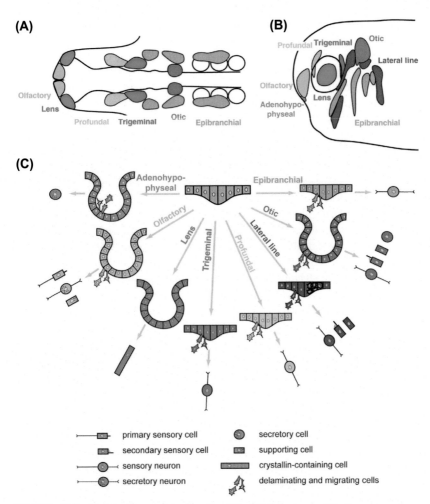

FIGURE 10.2 **Vertebrate cranial placodes.** (A) Chick embryo. (B) *Xenopus* embryo. (C) Developmental fates and derivative cell types of different cranial placodes. *(From Patthey, C., Schlosser, G., Shimeld, S.M., 2014. The evolutionary history of vertebrate cranial placodes–I: cell type evolution. Developmental Biology 389, 82–97.)*

between the adenohypophyseal placode and Hatschek's pit in amphioxus was discussed in Section 7.9 in connection with *Pitx* expression. Transcription factor genes, especially the *Eya* gene, three *Six* gene families (*Six3/6, Six1/2*, and *Six4/5*), and two *Pax* gene families (*Pax6* and *Pax2/5/8*) constitute a regulatory network in vertebrate placode development. In amphioxus, expression of many putative placode marker genes has been found in scattered ectodermal cells, presumed to comprise some or all of the sensory neurons described (Lacalli and Hou, 1999). Although it could be argued that these are homologous to vertebrate placode-derived sensory neurons at one level (ie, as surface ectoderm-derived sensory cells), the lack of focused domains of cells expressing these genes does not support homology at the level of a placode (Fig. 10.3A).

In contrast, many studies have suggested placode-homologous organs of ascidians because the aforementioned transcription factor genes were found expressed in the siphon primordium (Mazet et al., 2005; Fig. 10.3B). The ascidian *Pax2/5/8* gene, orthologous to the vertebrate otic placode markers, *Pax8* and *Pax2*, is expressed in the atrial primordium, supporting Jefferies' hypothesis about the homology of these structures (Wada et al., 1998). Moreover, *Pitx* expression in the oral siphon supports the oral siphon as an olfactory-adenohypophyseal placode homolog (Boorman and Shimeld, 2002; Christiaen et al., 2002; Fig.10.3C). Such studies have also suggested, although not very persuasively, that the ascidian palps (anterior secretory organs that lie just anterior to the oral siphon primordium) may also be olfactory homologs. This idea was partly dependent on early expression of *Eya* in cells that give rise to the palps and on identification of sensory neurons among the secretory cells (Mazet et al., 2005; Candiani et al., 2005).

In addition, recent studies reveal the presence of placode primordium in *C. intestinalis*. This species exhibits a proto-placodal ectoderm (PPE) that requires inhibition of BMP and that expresses the key regulatory determinant *Six1/2* and its cofactor *Eya*, a developmental process conserved across vertebrates, as mentioned earlier (Abitua et al., 2015). The *Ciona* PPE is shown to produce ciliated neurons that express genes for GnRH, a G-protein–coupled receptor for relaxin-3 and a functional cyclic nucleotide-gated channel, suggesting dual chemosensory and neurosecretory activities. These observations provide evidence that *Ciona* has a neurogenic protoplacode, which forms neurons that appear related to those derived from the olfactory placode and hypothalamic neurons of vertebrates, an ancestral cell type of vertebrates (Fig. 10.3B and D).

10.1.3 The Cardiopharyngeal Field

Although neural crest and placode play pivotal roles in formation of the new head, head muscles that are a crucial component of the complex vertebrate head are not produced by crest or placodes. In addition, another innovation of vertebrates is a chambered heart, which presumably allowed for the increased growth and metabolism demanded by active predation. These muscles originate from the cardiopharyngeal field (CPF) of the embryo (see review by Diogo et al., 2015 and related references therein).

The CPF is a developmental domain that gives rise to the heart and branchiomeric muscles. As shown in Fig. 10.4A, the amniote four-chambered heart is composed of cardiomyocytes derived from two adjacent progenitor cell populations in the early embryo (Meilhac et al., 2004). Early differentiating cardiac progenitor cells of the first heart field (FHF; red in Fig. 10.4A) give rise to the linear heart tube and later form the left ventricle and parts of the atria (Kelly, 2012).

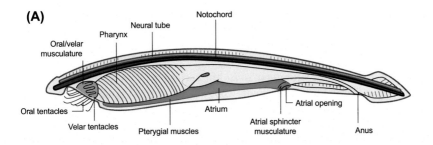

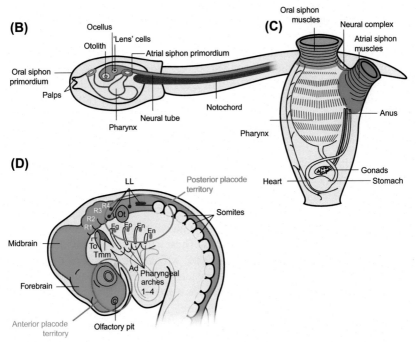

FIGURE 10.3 Hypotheses about homologies of placodes and branchiomeric muscles of chordates. (A) An adult cephalochordate showing the urochordate–cephalochordate muscle homology hypotheses proposed by Diogo et al. (2015). (B) Urochordate fish-like larva (anterior to the left). The notochord is in *red* and two siphon primordia are in *green* and *orange*, with putative relationships to the anterior and posterior placode territories shown in A. (C) An adult urochordate showing siphon primordia after metamorphosis. (D) Location of ectodermal placodes in the vertebrate head, according to Graham and Shimeld's hypothesis (anterior to the left): the olfactory placode or pit (*red*) at the tip of the forebrain; lens placodes (*orange*) form posteriorly as part of the eye; the adenohypophyseal placode (Ad, *yellow*) lies ventrally to the forebrain; trigeminal placodes form alongside the anterior hindbrain at the levels of rhombomeres 1 and 2 (R1 and R2), the anterior one being the ophthalmic placode (To, *light blue*) and the posterior one the maxillomandibular placode (Tmm, *purple*); otic placode (Ot, *brown*) forms opposite the central domain of hindbrain; lateral line placodes (LL, *pink*) form anteriorly and posteriorly to the otic placode; epibranchial placodes (*green*)—geniculate (Eg), petrosal (Ep), and nodose (En)—form as part of a pharyngeal series. Forebrain, midbrain, R1–4, and neural tube are shown in *dark blue*. (*From Diogo, R., Kelly, R.G., Christiaen, L., Levine, M., Ziermann, J.M., Molnar, J.L., Noden, D.M., Tzahor, E., 2015. A new heart for a new head in vertebrate cardiopharyngeal evolution. Nature 520, 466–473.*)

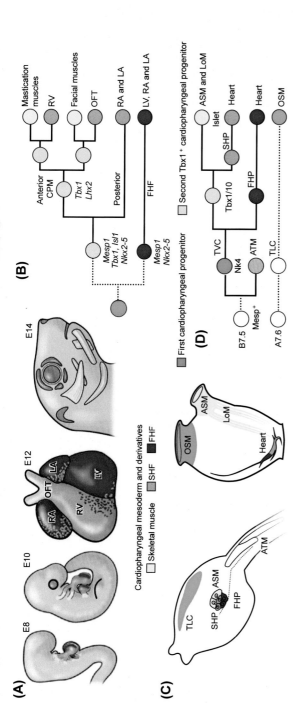

FIGURE 10.4   An evolutionarily conserved cardiopharyngeal ontogenetic motif. (A) Mouse embryos at embryonic days E8 and E10, the four-chambered mouse heart at E12, and the mouse head at E14. First heart field (FHF)-derived regions of heart [left ventricle (LV) and atrial are in *red*; second heart field (SHF)-derived regions of heart [right ventricle (RV), left atrium (LA), right atrium (RA), and outflow tract (OFT)] are in *yellow*; extraocular muscles are in *purple*. (B) A lineage tree depicting origins of cardiac compartments and branchiomeric skeletal muscles in mice. All cells derive from common pan-cardiopharyngeal progenitors (*dark green*) that produce the FHF, precursors of the LV and atria, and the second *Tbx1*+cardiopharyngeal progenitors (*light green*). Broken lines indicate that the early common FHF and SHF progenitor remains to be identified in mice. In anterior cardiopharyngeal mesoderm (CPM), progenitor cells activate Lhx2; self-renew; and produce the SHF-derived RV and OFT and first and second arch branchiomeric muscles (including muscles of mastication and facial expression). (C) Cardiopharyngeal precursors in *C. intestinalis* hatching larvae (left) and their derivatives in the metamorphosed juvenile (right). First heart precursors (FHP; *red*) and second heart precursors (SHP; *orange*) contribute to the heart (*red and orange mix*), whereas atrial siphon muscle precursors (ASM; *yellow*) form atrial siphon and longitudinal muscles (LoM; *yellow*). Oral siphon muscles (OSM; *blue*) derive from a heterogeneous larval population of trunk lateral cells (TLC, *blue*). ATM, anterior tail muscles. CPM is bilaterally symmetrical around the midline (*dotted line*). (D) Lineage tree depicting clonal relationships and gene activities deployed in *C. intestinalis* cardiopharyngeal precursors. All cells derive from *Mesp*+B7.5 blastomeres, which produce ATM (*gray*, see also left panel of C) and trunk ventral cells (TVC, *dark green*). The latter pan-cardiopharyngeal progenitors express *Nk4* and divide asymmetrically to produce the FHP (*red*) and second TVCs, the *Tbx1/10*+second cardiopharyngeal progenitors (second TVC, *light green disk*). The latter divide again asymmetrically to produce SHP (*orange*) and the precursors of ASM and LoM, which upregulate Islet. The OSM arise from A7.6-derived trunk lateral cells (TLC, *light blue*). (*From Diogo, R., Kelly, R.G., Christiaen, L., Levine, M., Ziermann, J.M., Molnar, J.L., Noden, D.M., Tzahor, E., 2015. A new heart for a new head in vertebrate cardiopharyngeal evolution. Nature 520, 466–473.*)

Subsequently, SHF progenitors (orange in Fig. 10.4A), located in pharyngeal mesoderm, produce cardiac muscle tissue (myocardium) of the outflow tract, right ventricle, and parts of the atria. The SHF can be divided into anterior and posterior progenitor cell populations that contribute to the arterial and venous poles of the heart, respectively. Cells from pharyngeal mesoderm can form either cardiac or skeletal muscle (*yellow* in Fig. 10.4A), depending on signals from adjacent pharyngeal endoderm, surface ectoderm, and neural crest cells (Nathan et al., 2008). The latter have important roles in regulating development of the CPF. They are required for deployment of SHF-derived cells to the heart's atrial pole. In addition, neural crest-derived mesenchyme patterns branchiomeric muscle formation (*yellow* in Fig. 10.4A) and gives rise to associated fascia and tendons.

The lineages of cells involved in the CPF have been traced with specific expression of transcription factor genes. All cells derive from common pan-cardiopharyngeal progenitors (*dark green* in Fig. 10.4B). The progenitors produce *Mesp+/Nkx2-5* + FHF precursors (*red*) of the left ventricle and atria and the second *Tbx1* + cardiopharyngeal progenitors (*light green*). Then, in anterior cardiopharyngeal mesoderm, progenitor cells activate *Lhx2*, self-renew, and produce the SHF-derived right ventricle and the outflow tract (*orange*) and first and second arch branchiomeric muscles (*yellow*; including muscles for mastication and facial control).

Interestingly, a similar pattern of muscle development exists in *C. intestinalis* (Fig. 10.4C). In *Ciona*, the FHF and SHF progenitors contribute to the heart whereas atrial siphon muscle precursors form the atrial siphon and longitudinal muscles. On the other hand, oral siphon muscles derive from a heterogeneous larval population of trunk lateral cells. The cardiopharyngeal mesoderm of *Ciona* larvae is bilaterally symmetrical around the midline (*dotted line* in Fig. 10.4C left). Lineage tracing experiments show that the B7.5 blastomere of the 64-cell–stage embryo contributes to FHF progenitors, SHF progenitors, atrial siphon muscles, and longitudinal muscles (Fig. 10.4D). All cells derived from *Mesp* + B7.5 produce atrial siphon muscles and trunk ventral cells (TVCs). The latter pan-cardiopharyngeal progenitors express *Nk4* and divide asymmetrically to produce the FHF progenitors and second TVCs. The *Tbx1/10* + second cardiopharyngeal progenitors (second TVCs) divide again asymmetrically to produce second heart precursors (SHP) and precursors of atrial siphon muscles and longitudinal muscles, which upregulate *Islet* (Fig. 10.4D). On the other hand, the oral siphon muscles arise from A7.6-derived trunk lateral cells (Fig. 10.4D).

10.1.4 Cartilage and Bone

Vertebrate cartilage and bone are used for protection, predation, and endoskeletal support. Because there are no similar tissues in cephalochordates or urochordates, these tissues represent a major leap in vertebrate evolution (eg, Kardong,

2014). It appears that mineralized tissues developed gradually during vertebrate evolution because extant jawless vertebrates (lamprey and hagfish) have no mineralized tissues. The earliest mineralized tissue was found in the feeding apparatus of extinct jawless fishes, the conodonts (see Section 3.4.3). Cartilaginous fish produce calcified cartilage and dermal bone, including teeth, dermal denticles, and fin spines, but their cartilage is not replaced with endochondral bone. Endochondral ossification is accomplished by a highly complicated process unique to bony vertebrates. Recent decoding of the elephant shark genome suggests that the lack of genes encoding secreted calcium-binding phosphoproteins in cartilaginous fishes explains the absence of bone in its endoskeleton (Venkatesh et al., 2014).

In vertebrates, consistent with their structural roles, there may be a direct relationship between notochord and cartilage because both express many of the same genes, such as those that encode type II and type IX collagen, aggrecan, Sox9, and chondromodulin. During endochondral bone formation, the type II collagen-rich extracellular matrix of cartilage is formed with type X collagen, which signals the eventual replacement of cartilage by bone (Linsenmayer et al., 1986; Schmid et al., 1991; Solursh et al., 1986). Likewise, during the development of vertebrae, the notochord that runs through the middle of each vertebra first expresses type X collagen and is then replaced by bone (Linsenmayer et al., 1986). Between the vertebrae, the notochord does not express type X collagen and is not replaced by bone but becomes the center of the intervertebral disc—the nucleus pulposus (Aszódi et al., 1998; Smits and Lefebvre, 2003). Thus notochord can become ossified in a fashion similar to cartilage. Consistent with this view, in mutant mice that lack type II collagen the notochord is not replaced by bone, presumably because the type II collagen network is required for proper deposition of type X collagen.

10.1.5 Adaptive Immune System

All metazoans protect themselves against pathogens using sophisticated immune systems. Immune responses of invertebrates are innate and usually stereotyped. On the other hand, vertebrates adopted an additional system of adaptive immunity that uses immunoglobulins, T-cell receptors, and major histocompatibility complex (MHC) molecules (Boehm, 2012). The adaptive immune system enables more rapid and efficient response upon repeated encounters with a given pathogen. Surveys of cephalochordate and urochordate genomes failed to detect genes encoding immunoglobulins, T-cell receptors, or MHC molecules, indicating that the adaptive immune system is another vertebrate innovation (Azumi et al., 2003; Holland et al., 2008). Recent discoveries of alternative antigen receptor systems in jawless vertebrates suggest that cellular and molecular changes involved in evolution of the vertebrate adaptive immune system are more complex than previously thought (Pancer et al., 2004; Guo et al., 2009).

10.2 EVOLUTION OF UROCHORDATES

More than 520 MYA, a chordate stem that gave rise to tunicates incorporated horizontally transferred bacterial genes for cellulose synthesis into the genome (see Section 11.3.1). The acquisition of a cellulose-based tunic as an outer shield for the body was likely deeply involved in alteration of their lifestyles (Satoh, 2009). Unlike ancestral, free-living adults, they adopted an immobile lifestyle by adhering strongly to rocks and other substrates. Viewed from the standpoint of energetics, this sessile lifestyle requires much less energy compared with active predation. To accomplish this adult style, developmental processes were greatly shortened to quickly produce larvae. These larvae do not open their mouths until they settle and complete metamorphosis. An interesting point is that although tunicates are highly derived (diversified) from the mainstream that leads to vertebrates, they share more traits with vertebrates than with cephalochordates. This supports tunicates as a sister group of vertebrates, as suggested by molecular phylogeny. Various innovations found in tunicates have already been discussed in my previous book (Satoh, 2014). Here, only four issues will be briefly described.

10.2.1 Precocious Mode of Embryonic Development

Oikopleura dioica, the most widely used larvacean and laboratory model, has a very short life cycle (4 days at 20°C); only 12 h are required for embryogenesis and approximately 4 days for planktonic filter-feeders to capture small particles with their gelatinous houses. This is one example of precocious embryonic development. In contrast to other deuterostome groups, cleavage of ascidians is bilaterally symmetrical. Gastrulation begins at approximately the 110-cell stage, and a newly hatched larva consists of only 2600 cells with typical cell types of the chordate body. The developmental mode is mosaic and developmental fate is determined very early in embryogenesis including maternal muscle determinant *Macho-1* (Nishida and Sawada, 2001); the basic blueprint for future chordate-type body is established as early as the 110-cell stage (Imai et al., 2006). Taking advantage of a well-described lineage and a great quantity of transcriptomic data, expression and function of genes encoding transcription factors and signal pathway molecules have been detailed. GRNs in ascidian embryos appear to be much simpler than those of other chordates (Satoh, 2014).

10.2.2 Asexual Reproduction of Colonial Ascidians

Despite being the chordates closest to vertebrates, urochordates have a colonial lifestyle that evolved independently in several lineages of ascidians. Colonial ascidians such as *Botryllus*, *Botrylloides*, and *Polyandrocarpa* reproduce sexually and asexually by budding and by strobilation (Fig. 10.5A). Extensive diversity exists with respect to modes of bud formation. However, regardless of the mode of budding, the oozoid (developed from an egg) usually reproduces only asexually and does not attain sexual maturity. On the other hand, the blastozoid (developed from a bud)

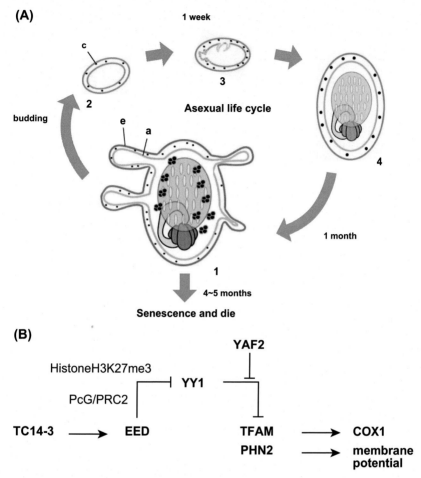

FIGURE 10.5   Budding of colonial ascidians and the underlying mechanism. (A) Asexual reproduction and aging of *P. misakiensis*. (B) Molecular network to maintain the stem cell-like condition of endoderm cells. *COX1*, cytochrome oxidase subunit 1; *EED*, H3K27 methyltransferase partner; *PHN2*, Prohibitin 2; *TFAM*, mitochondrial transcription factor; *YY1*, Yin Yang1; *YAF2*, YY1-associated factor 2. *(Courtesy of Kazuo Kawamura.)*

reproduces sexually and asexually. In other words, without at least one phase of asexual development, colonial ascidians would not propagate sexually, suggesting that asexual reproduction is an adaptation for propagation in colonial ascidians.

10.2.3 Pluripotent Stem-Like Cells for Bud Formation

Modes of bud formation can be categorized based on cell types from which major organs of the bud are derived. That is, although buds originate from the parent body, not every cell type constituting the parent body is necessarily incorporated into buds. Therefore unincorporated cell types are supplied by multipotent cells that exist in the bud. *Polyandrocarpa misakiensis* adopts peribranchial budding, and buds are formed via protrusions of the parent zooid body wall. At Kochi University, *P. misakiensis* has been maintained as a clone by only asexual budding for more than 25 years (>500 generations; Kazuo Kawamura, personal information). However, adults die 4–5 months after starting bud formation (Fig. 10.5B). This raises questions about how stem-like cells for budding maintain multipotency and how senescence occurs in the adult body. The answer to these questions may provide insights into how certain lineages of tunicates gained such ability and what kinds of molecular mechanisms are involved in the process. Interestingly, epigenetic control mechanisms are very likely involved (Fig. 10.5B; Kawamura et al., 2015).

10.2.4 The Development of Colony Specificity

When two pieces from a single colony are placed in juxtaposition at natural growing edges or artificially cut surfaces, they fuse to form a common vascular system. This phenomenon is called fusion. By contrast, juxtaposition of tissues from different colonies usually results in necrosis. This phenomenon is termed the nonfusion reaction, or simply, rejection. These phenomena collectively define colony specificity and are observed in most colonial ascidians. Colony specificity has attracted attention because it offers significant insights into the origin of the allorecognition histocompatibility of vertebrates. Recent decoding of the genome of *Botryllus schlosseri* and related experiments disclosed a gene that encodes a protein with highly modified sugar chains (Voskoboynik et al., 2013).

10.3 CONCLUSION

After their appearance, chordates diversified into three separate lineages. Extant lancelet cephalochordates most likely kept their original (ancestral) form, evident in fossils and suggesting a good match of body plan to lifestyle. On the other hand, urochordates and vertebrates have changed more dynamically to adapt to their lifestyles, the former becoming advanced filter feeders whereas the latter became predators. The evolutionary alteration of vertebrate anatomy and physiology is tremendous, and various innovations took place to form a new body plan with an elaborated head, sensory system, chambered heart, etc. These evolutionary developmental events are also interpreted in the new organizers hypothesis.

Chapter 11

How Did Chordates Originate and Evolve?

I have discussed the evolutionary developmental mechanisms involved in the origins and evolution of chordates. This chapter further discusses this problem from a broader point of view. Ignoring detailed genetic changes, how did chordates appear and diversify? Here I will discuss how novel structures and functions emerged. It is generally accepted that morphological structures of various multicellular animals are based on a common set of toolkit genes and that novelty results from altered regulation of toolkit gene expression. It is also apparent that new or modified structural genes have contributed to the appearance of novel structures.

11.1 THE THREE-PHYLUM SYSTEM OF CHORDATES

Traditionally, deuterostomes have been thought to comprise three phyla: Echinodermata, Hemichordata, and Chordata (Fig. 11.1A). The Chordata consists of three subphyla: Cephalochordata, Urochordata, and Vertebrata (eg, Stach, 2008). Characteristic features of the five deuterostome taxa have been repeatedly discussed. Indeed, these five diverged approximately 570 million years ago (MYA) or earlier and have undergone their own evolutionary processes. Do recent data support this phylogenetic classification, and does it accurately reflect deuterostome relationships? It is now the general consensus that the Echinodermata and Hemichordata comprise a clade known as the Ambulacraria and that the Chordata includes the Cephalochordata, the Urochordata (Tunicata), and the Vertebrata. Here, I propose a three-phylum system to replace the current three-subphylum system of chordates (Fig. 11.1B) (Satoh et al., 2014a). These are two major reasons for this change.

First, using molecular phylogenetic techniques, protostomes have now been reclassified into two major, monophyletic groups above the phylum level: the Lophotrochozoa (perhaps Spiralia in general) and the Ecdysozoa. These two are readily distinguishable by their different developmental pathways. The former is characterized by spiral cleavage and the latter by exoskeleton molting. With robust support from molecular phylogenomics and evolutionary developmental biology, the deep gap between the Chordata (fish-like larvae) and the Ambulacraria (dipleurula larvae) merits a taxonomic classification above the phylum. Thus the Lophotrochozoa (consisting of ~15 phyla), the Ecdysozoa (~8 phyla),

(A) current view

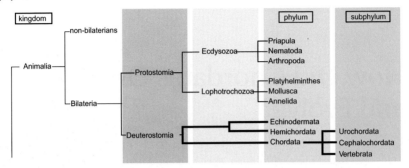

(B) a proposed view

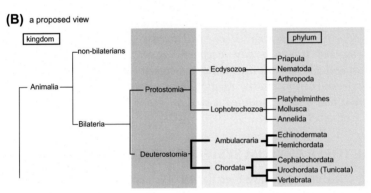

FIGURE 11.1 A traditional, current view (A) and the proposed view (B) of chordate phylogeny with respect to interphylum relationships. The proposed phylogeny regards the Cephalochordata, the Urochordata, and the Vertebrata as separate phyla, rather than as subphyla. (*From Satoh, N., Rokhsar, D., Nishikawa, T., 2014. Chordate evolution and the three-phylum system. Proceedings of the Royal Society B: Biological Sciences 281, 20141729.*)

the Ambulacraria (2 phyla), and the Chordata (3 phyla) should all be classified at the superphylum level (Fig. 11.1B). As emphasized in this book, the occurrence of fish-like larvae is fundamental to chordate origins, and the three chordate taxa each have characters distinct enough to be treated as phyla.

Second, metazoans are classified into approximately 34 phyla. However, only a few are distinguished by specific, diagnostic structures, such as nematocytes (Cnidaria), comb plates (Ctenophora), or segmented appendages (Arthropoda). Consistent with this classification at least the Urochordata and Vertebrata have unique structural features, supporting their recognition as phyla. Characteristic features of the three phyla are as follows.

Cephalochordates or lancelets comprise only approximately 35 species of small, fish-like animals that burrow in sand. They are often called "acraniates" because their central nervous system (CNS) consists of a neural tube with a small anterior vesicle that does not develop into the three-part brain seen in urochordate larvae and vertebrates. These suspension feeders develop a well-organized feeding and digestive system, a ciliary-mucous system having a wheel organ, Hatschek's pit, an endostyle, and a pharynx with gill slits. In addition, vertebrate-like myotomes that develop from larval somites facilitate very rapid, fish-like locomotion. The lancelet notochord has muscle-like properties.

Urochordates comprise approximately 3000 extant species with a great variety of lifestyles, including sessile ascidians, planktonic and tadpole-like juvenile larvaceans, and planktonic, barrel-shaped thaliaceans. A distinctive feature that characterizes urochordates as a phylum is that they are the only animal group that can directly synthesize cellulose (see discussion in Section 11.3.1). They appear to have evolved as filter-feeding specialists.

Vertebrates are the most specialized animal taxa and have distinctive features that are not found in other metazoans. These include a neural crest, a placode, an endoskeleton, an adaptive immune system, etc., as discussed in Chapter 10. The present reclassification of chordate groups provides a better representation of their evolutionary relationships, which will benefit future studies of chordate and vertebrate origins.

11.2 MECHANISMS INVOLVED IN ORIGINATION OF DEUTEROSTOME NOVELTIES

The question of how novel structures develop as a result of changes in genetic information is one of the most challenging issues in evolutionary developmental biology. It is generally accepted that morphological features of multicellular animals are based on a common set of transcription factor genes and signaling pathway molecules, and that novel features emerge as a result of modifications to regulatory networks of these genetic toolkit genes [gene regulatory networks (GRNs); Carroll et al., 2005; Davidson, 2006; Peter and Davidson, 2015]. Novel genetic material has also contributed to the evolution of new structures, and considerable attention has been paid to gene duplications as a source of new genetic material. In the case of vertebrate evolution, two rounds of genome-wide gene duplication (2R-GWGD) are regarded as the main events driving evolution of novel structures. Domain shuffling is another mechanism to create new proteins, and several different molecular mechanisms for domain shuffling have been proposed. Because domains are often correlated with exon boundaries, exon shuffling is believed to be a driving force behind domain shuffling. Here I discuss examples of genetic changes in relation to deuterostome evolution.

11.3 HORIZONTAL GENE TRANSFER

Horizontal gene transfer (HGT), also known as lateral gene transfer, refers to the movement of genetic information between more or less distantly related organisms; thus it has nothing to do with the standard vertical transmission of genes from parent to offspring. HGT is accepted as an important evolutionary force often modulating the evolution of prokaryote genomes, and sometimes eukaryote genomes as well. Although it has been thought that HGT plays only a minor role in metazoan evolution, recent studies have demonstrated an increasing number of examples of HGT in metazoans (eg, Graham et al., 2008; reviewed by Boto, 2014), including sponges, cnidarians, rotifers, nematodes, molluscs, and arthropods. For example, genomic analysis of the rotifer, *Adineta vaga*, reveals the presence of 8% foreign genes (Flot et al., 2013). The presence of the cellulose synthase gene (*CesA*) in the urochordate genome is the best example of HGT in metazoans. In addition, a recent study shows that genes encoding enzymes involved in sialic acid biosynthesis are a deuterostome novelty resulting from HGT (Simakov et al., 2015).

11.3.1 Acquisition of Cellulose Biosynthetic Ability and Urochordate Evolution

Urochordates are the only animals that can synthesize cellulose, a biological function normally associated with bacteria and plants but not metazoans. As first noticed in the early 19th century (eg, Lamarck, 1816), the entire adult urochordate body is invested with a thick covering called the tunic or test (Fig. 11.2A); hence, the common name, tunicates. The tunic may function as an outer protective structure, similar to a mollusc shell, and has undoubtedly influenced the evolution of lifestyles within this group (see Section 10.2). As discussed in Chapter 3, Early Cambrian fossil tunicates from South China, such as *Cheungkongella* (Shu et al., 2001; Fig. 3.4D) and *Shankou*, exhibit an outer morphology similar to that of extant ascidians, suggesting that ascidians already had a tunic as early as 520 MYA. A major constituent of the tunic is tunicin, a type of cellulose (Fig. 11.2B).

Cellulose is a polysaccharide consisting of linear chains of several thousand $\beta(1\rightarrow4)$-linked D-glucose units and it is the most abundant biological molecule on Earth. Two major components of cellulose production are cellulose synthases (CesA; membrane-embedded glycosyltransferases) and cellulase [a family of cellulose dehydrogenases (CDs)]. Prokaryotic and eukaryotic CesAs share a similar predicted topology including eight transmembrane helices and at least one extended intracellular glycosyltransferase (GT) loop. CDs have three major

FIGURE 11.2 **Cellulose biosynthesis in tunicates.** (A) An adult ascidian, *Halocynthia roretzi*. (B) Treatment of an adult with sodium hydroxide showing that adults are covered with a cellulosic tunic. (C) Staining of cellulose on the outer surface of *C. intestinalis* larva. (D) Larva of a mutant called "swimming juvenile (*sj*)" showing the lack of cellulose formation. *(Courtesy of Keisuke Nakashima and Yasunori Sasakura.)* (E) Schematic representation of the *Ci-CesA* gene displays the total 14-kb region of two Scaffolds, which predict two genes, ci0100130874 and ci0100152699. Sequence comparison of the Scaffold sequences with the full-length *Ci-CesA* cDNA reveals the intron-exon structure of *Ci-CesA*. *Ci-CesA* contains 21 exons. (F) Comparison of the domain structure of cellulose synthases of various origins. Multiple alignments are shown in schematic drawings indicating the positions of the common motifs, depicted as D, DxD, D, and QxxRW. Only the cellulose synthase domain of Ci-CesA is represented. In comparison with ascidian and prokaryotic sequences, algal and plant sequences conserved an N-terminal zinc-binding domain and two insertions, the so-called plant-specific conserved region (CR-P) and a class-specific region, depicted in *gray*. *(From Nakashima, K., Yamada, L., Satou, Y., Azuma, J., Satoh, N., 2004. The evolutionary origin of animal cellulose synthase. Development Genes and Evolution 214, 81–88.)* (G, H) Molecular phylogeny based on the comparison of amino acid sequences of (G) CesA domains and (H) GH6 domain. *(Courtesy of Keisuke Nakashima.)*

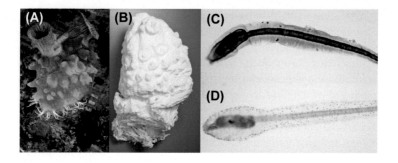

(E)

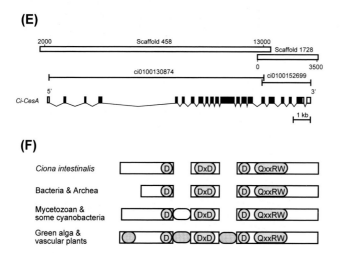

(F)

(G) CesA   **(H)** GH6

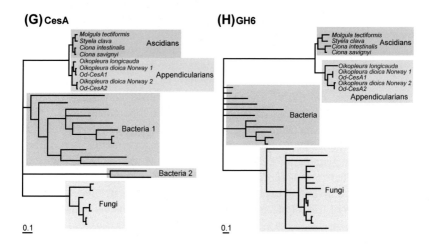

types: CD6, CD8, and CD9. Most bacteria, except cyanobacteria, have cellulase with CD8. *Streptomyces* has CD6, and most metazoans have CD9 for digestion of cellulose in foods. In plants, cellulose is synthesized by a large, oligomeric protein complex in the plasma membrane called the terminal complex, whereas in bacteria, cellulose synthesis and transport across the inner bacterial membrane are mediated by the CesA complex (Morgan et al., 2013).

The possibility of HGT in tunicates has received support from the identification and characterization of a single-copy of *CesA* in both the *Ciona intestinalis* and *Ciona savignyi* genomes (Nakashima et al., 2004; Matthysse et al., 2004). *Ci-CesA* (*CesA* of *C. intestinalis*) is composed of 21 exons comprising approximately 14 kb of the genome (Fig. 11.2E). It encodes a transmembrane protein with 1277 amino acid residues. Interestingly, Ci-CesA is a fusion of a cellulose synthase and a cellulase because it contains a β-glycosyltransferase (GT) of the GT-2 family and a glycoside hydrolase (GH) of the GH-6 family (Fig. 11.2F). Such atypical fusion genes are not found in public databases, suggesting that the two domains originated from two distinct genes. Molecular phylogeny using GT-6s suggests that Ci-CesA belongs to a clade with *Streptomyces* CesA, suggesting that the bacterial *CesA* was transferred into the genome of a tunicate ancestor more than 550 MYA (Fig. 11.2G and H). *Ci-CesA* is expressed in larval and adult epidermis (Fig. 11.2C). Its function in cellulose biosynthesis became evident from a mutant called *swimming juvenile* (*sj*), in which the enhancer element of *Ci-CesA* is a transposon-mediated mutation and thereby lacks cellulose biosynthetic activity (Fig. 11.2D; Sasakura et al., 2005).

In addition, tunicates display another interesting biochemical phenomenon in relation to cellulose. The native form of cellulose is a fibrillar composite of two crystalline phases—the triclinic I_α and monoclinic I_β allomorphs—and allomorph ratios are specific to species. Cellulose produced by bacteria and algae is enriched in I_α whereas cellulose of higher plants consists mainly of I_β. However, the mechanisms responsible for crystal formation remain unknown. In the larvacean, *Oikopleura dioica*, *CesA* is duplicated into *Od-CesA1* and *Od-CesA2* (Sagane et al., 2010). *Od-CesA1* encodes larval cellulose whereas *Od-CesA2* encodes adult cellulose that frames a mucous filter-feeding device, the "house." Knockdown of *Od-CesA1* resulted in the failure of cellulose biosynthesis.

Interestingly, *Od-CesA1* encodes larval cellulose consisting of the I_α allomorph whereas *Od-CesA2* encodes adult cellulose consisting of the I_β allomorph (Nakashima et al., 2011). Both structures are secreted from the epidermis under the mutually exclusive expression patterns of the two cellulose synthase genes. Namely, in *O. dioica*, an original *CesA* was duplicated into genes under differential control to produce different forms of cellulose. This provides an experimental system to explore molecular mechanisms involved in differential production of triclinic I_α and monoclinic I_β allomorphs via transcription of corresponding genes.

Although it is evident that tunicates gained *CesA* by HGT, this evolutionary event raises various questions as to how the transferred gene became functional.

Not only did the gene jump into the ancestral tunicate genome, but it also acquired the ability to be expressed in epidermal cells of developing embryos. How did the gene come under the control of epidermis-specific expression? This and other questions should be explored in future studies. In addition, initially, how did tunicates benefit from their ability to synthesize cellulose? The tunic may function as an outer protective structure, similar to a mollusc shell. Compared with other invertebrate deuterostomes, tunicates show highly divergent lifestyles, including asexual reproduction, discussed in Section 10.2.

11.3.2 Genes Encoding Enzymes Involved in Sialic Acid Metabolism

Hemichordate genome analysis identified more than 30 genes related to functional innovation in deuterostomes, having markedly different sequences from those of other metazoans. Some may have arisen from accelerated sequence change in the deuterostome stem lineage, originating from distant, but identifiable bilaterian homologs. Others represent novel deuterostome protein domain combinations, whereas others lack identifiable sequence and domain homologs in other metazoans. However, the latter group comprises more than a dozen deuterostome genes that have readily identifiable homologs among marine microbes, often cyanobacteria, or eukaryotic microalgae.

Regardless of their origins, various deuterostome novelties and gene family expansions relative to sialic acid metabolism are noteworthy (Kondrashov et al., 2006; Simakov et al., 2015). Deuterostomes are unique among metazoans in their high level and diverse covalent linkages of sialic acid (also known as *N*-acetyl neuraminic acid), a nine-carbon, negatively charged sugar, to the terminal sugars of glycoproteins, mucins, and glycolipids. Hemichordates contain expanded families of enzymes for several of these reactions (Fig. 11.3). Judging from the presence/absence of relevant enzymes, 5 of the 11 reactions involved in sialic acid metabolism are not found in protostomes or other metazoans (Fig. 11.3). Because the other six steps use enzymes similar to those of some protostomes, it is highly likely that these five are deuterostome novelties.

The importance of glycoproteins for mucociliary feeding and other hemichordate activities is further supported by the novel and expanded families of genes encoding the polypeptide backbones of glycoproteins, such as those with von Willebrand type-D and/or cysteine-rich domains, including mucins. Mucin genes are present in hemichordates and amphioxus as large, tandemly duplicated clusters with varied expression patterns. These are not found in the sea urchin genome because echinoderms have a different mode of feeding. In hemichordates and amphioxus, the pharynx is heavily ciliated. Cells of the pharyngeal walls in hemichordates and the ventral endostyle in amphioxus secrete abundant mucins and glycoproteins. Similarly, in the deuterostome ancestor these glycoproteins probably enhanced the mucociliary filter-feeding capture of food particles from the microbe-rich marine environment and

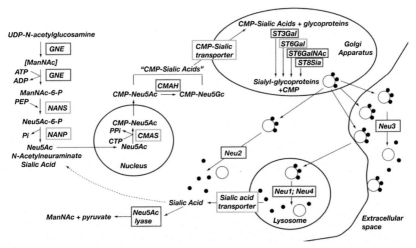

FIGURE 11.3   Gene family expansion of deuterostome enzymes involved in sialic acid metabolism. Biosynthesis of sialic acid and its addition to and removal from glycoproteins are shown. Although many steps involve enzymes similar to those of the more limited pathway of some protostomes, deuterostomes are unique among metazoans in their high level and diverse linkages of sialic acid (also known as neuraminic acid), a nine-carbon, negatively charged sugar, as the terminal sugar of carbohydrate chains of glycoproteins, mucins, and glycolipids. There are expanded families of enzymes for several of these reactions in hemichordates. *Red boxes* around protein names indicate deuterostome novelties. (*From Simakov, O., Kawashima, T., Marlétaz, F., Jenkins, J., Koyanagi, R., Mitros, T., Hisata, K., Bredeson, J., Shoguchi, E., Gyoja, F., Yue, J.X., Chen, Y.C., Freeman, R.M., Sasaki, A., Hikosaka-Katayama, T., Sato, A., Fujie, M., Baughman, K.W., Levine, J., Gonzalez, P., Cameron, C., Fritzenwanker, J.H., Pani, A.M., Goto, H., Kanda, M., Arakaki, N., Yamasaki, S., Qu, J., Cree, A., Ding, Y., Dinh, H.H., Dugan, S., Holder, M., Jhangiani, S.N., Kovar, C.L., Lee, S.L., Lewis, L.R., Morton, D., Nazareth, L.V., Okwuonu, G., Santibanez, J., Chen, R., Richards, S., Muzny, D.M., Gillis, A., Peshkin, L., Wu, M., Humphreys, T., Su, Y.H., Putnam, N.H., Schmutz, J., Fujiyama, A., Yu, J.K., Tagawa, K., Worley, K.C., Gibbs, R.A., Kirschner, M.W., Lowe, C.J., Satoh, N., Rokhsar, D.S., Gerhart, J., 2015. Hemichordate genomes and deuterostome origins. Nature 527, 459–465.*)

protected its inner and outer tissue surfaces in relation to the pharyngeal gene cluster (Chapter 6).

11.4 THE SIGNIFICANCE OF GENE DUPLICATION IN DEUTEROSTOME EVOLUTION

In addition to genome-wide gene duplication that occurred in the vertebrate lineage, duplication of genes, especially those encoding transcription factors and signal pathway molecules, plays pivotal roles in metazoan evolution. One example here is a deuterostome-specific novelty in the transforming growth factor-β (TGFβ) signaling pathway.

The signaling ligands Lefty (a Nodal antagonist) and Univin/Vg1/GDF147 (a Nodal agonist) are deuterostome innovations that modulate Nodal signaling during the major developmental events of endomesoderm induction and axial patterning in vertebrates, axial patterning in hemichordates and echinoderms,

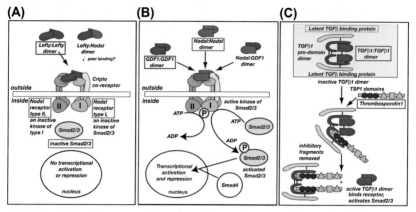

FIGURE 11.4   Novel genes in transforming growth factor (TGF)-β signaling pathways in deuterostomes. Three examples of novel genes in TGFβ pathways are shown. (A) Lefty, an antagonist of Nodal signaling, which activates Smad2/3-dependent transcription when not antagonized. (B) Univin, an agonist of Nodal signaling, also called Vg1, DVR1, and GDF1. (C) TGFβ2, a ligand that activates Smad2/3-dependent transcription by binding to a deuterostome-specific TGFβ receptor type II, which contains a novel ectodomain (not shown). Also shown in (C) is the novel protein, thrombospondin (TSP)-1, which activates TGFβ2 by releasing it from an inactive complex by way of its TSP1 domains. *Red boxes* around protein names indicate deuterostome novelties. *Green boxes* around names indicate genes with pan-metazoan/bilaterian ancestry and without accelerated sequence change in the deuterostome lineage. (*From Simakov, O., Kawashima, T., Marlétaz, F., Jenkins, J., Koyanagi, R., Mitros, T., Hisata, K., Bredeson, J., Shoguchi, E., Gyoja, F., Yue, J.X., Chen, Y.C., Freeman, R.M., Sasaki, A., Hikosaka-Katayama, T., Sato, A., Fujie, M., Baughman, K.W., Levine, J., Gonzalez, P., Cameron, C., Fritzenwanker, J.H., Pani, A.M., Goto, H., Kanda, M., Arakaki, N., Yamasaki, S., Qu, J., Cree, A., Ding, Y., Dinh, H.H., Dugan, S., Holder, M., Jhangiani, S.N., Kovar, C.L., Lee, S.L., Lewis, L.R., Morton, D., Nazareth, L.V., Okwuonu, G., Santibanez, J., Chen, R., Richards, S., Muzny, D.M., Gillis, A., Peshkin, L., Wu, M., Humphreys, T., Su, Y.H., Putnam, N.H., Schmutz, J., Fujiyama, A., Yu, J.K., Tagawa, K., Worley, K.C., Gibbs, R.A., Kirschner, M.W., Lowe, C.J., Satoh, N., Rokhsar, D.S., Gerhart, J., 2015. Hemichordate genomes and deuterostome origins. Nature 527, 459–465.*)

and left–right patterning in all deuterostomes (Fig. 11.4). *Univin* is tightly linked to the related bilaterian *bmp2/4* in the sea urchin genome (Fig. 5.4A). In hemichordates and amphioxus, its origin by tandem duplication and divergence from an ancestral *bmp2/4*-type gene has been suggested. Vertebrates contain TGFβ 1, 2, and 3. TGFβ2 signaling is a deuterostome innovation that controls cell growth, proliferation, differentiation, and apoptosis at later developmental stages (Fig. 11.4B). Accompanying the novel TGFβ2 ligand, the type II receptor has a novel ectodomain. The extracellular matrix protein thrombospondin-1 (TSP1), which activates TGFβ2 in vertebrates, contains a combination of domains that is unique to deuterostomes, including three TSP1 domains that bind the TGFβ2 prodomain region (Fig. 11.4C). Although these signaling novelties have clear sequence similarities to pan-bilaterian components, they form long-stem branch clades on phylogenetic trees, indicating extensive sequence divergence along the deuterostome stem. Together, these specific innovations appear to contribute to increased and complex patterning of Smad2/3-mediated signaling in deuterostomes compared with protostomes and other metazoans.

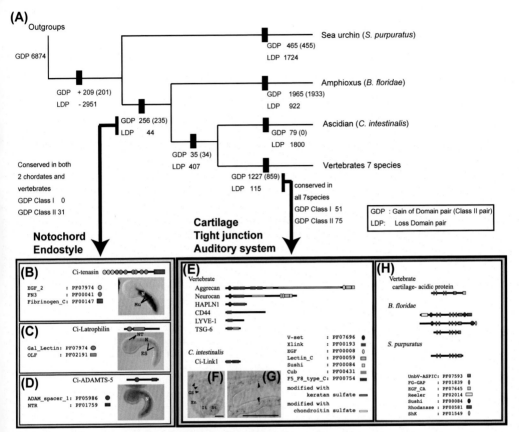

FIGURE 11.5   **Schematic showing the significance of domain shuffling during chordate evolution.** (A) Gain (GDP) and loss (LDP) of domain pairs are mapped on the deuterostome phylogenetic tree. Numbers on each node indicate GDP and LDP. Acquisitions of Class II domain pairs are shown in *brackets. Arrows* indicate domain pairs acquired by ancestral chordates and those in ancestral vertebrates. Among 1227 pairs acquired by ancestral vertebrates, 51 Class I domain pairs and 75 Class II domain pairs are conserved in 7 vertebrate species (middle). Some of these are involved in vertebrate-specific characters such as cartilage, tight junctions, and auditory systems. Among 256 domain pairs unique to chordates, 31 pairs are conserved in ascidians, lancelets, and various vertebrate species. Some of them are involved in chordate-specific characters such as the notochord and the endostyle. (B–D) Tenascin, latrophilin, and ADAMTS-5, which were created by domain shuffling in ancestral chordates, are expressed in notochords of ascidian embryos. Symbols indicated the domain ID and Pfam accession numbers. (B) *Ci-tenascin* is expressed in ascidian notochord (*N*) and muscle cells (*Mu*). (C) *Ci-latrophilin* is expressed in the notochord, neural tube (*NT*), and endodermal strand (*ES*) of ascidian larvae. (D) *Ci-ADAMTS-5* is expressed in ascidian notochord. (E) Domain structures of proteins that include an Xlink domain. (F, G) *Ci-link1* is expressed in some blood cells (*arrowheads*) in juvenile ascidians. Scale bars, 50 μm. *En*, Endostyle; *GS*, gill slits; *St*, stomach; *It*, intestine. (H) Domain structures of cartilage acidic protein and proteins containing an ASPIC-and-UnbV domain encoded by amphioxus gene models. (*From Kawashima, T., Kawashima, S., Tanaka, C., Murai, M., Yoneda, M., Putnam, N.H., Rokhsar, D.S., Kanehisa, M., Satoh, N., Wada, H., 2009. Domain shuffling and the evolution of vertebrates. Genome Research 19, 1393–1403.*)

11.5 SIGNIFICANCE OF DOMAIN SHUFFLING IN CHORDATE EVOLUTION

Comparative genomics of deuterostomes highlighted the significance of domain shuffling as one of the mechanisms that contributes to the evolution of vertebrate- and chordate-specific characteristics (Fig. 11.5). For example, Kawashima et al. (2009) identified approximately 1000 new domain pairs in the vertebrate lineage, including approximately 100 that are shared by all 7 of the vertebrate species examined. Some of these pairs occur in the protein components of vertebrate-specific structures, such as cartilage and the inner ear, suggesting that domain shuffling made a marked contribution to the evolution of vertebrate-specific characteristics (Fig. 11.5E–H). The evolutionary history of domain pairs is traceable. For example, the Xlink domain of aggrecan, one of the major components of cartilage, was originally utilized as a functional domain of a surface molecule of blood cells in chordate ancestors, and it was recruited by the protein of the matrix component of cartilage in the vertebrate ancestor (Fig. 11.5E). Some of the genes that were created as a result of domain shuffling in ancestral chordates are involved in functions of chordate structures, such as the endostyle, Reissner's fiber of the neural tube, and the notochord (Fig. 11.5B–D).

11.6 THE SIGNIFICANCE OF STRUCTURAL GENES IN METAZOAN EVOLUTION

The expression and function of genes that encode transcription factors and signal pathway molecules and that play pivotal roles in development and evolution of metazoan body plans are evident. GRNs and modifications of components in GRNs are key to understanding genetic mechanisms of metazoan evolution. However, the evolution of animal body plans is much more complex than we expect, and studies of GRNs do not always disclose the mechanisms. Much more attention should be paid to the significance of structural genes downstream of GRNs and the combinatorial performance of these two.

One example is the role of *Pax6* in eye formation in metazoans (Gehring, 2005; Erclik et al., 2009). Vertebrates develop camera (lens) eyes in combinatorial control of the neural placode and the nervous system whereas *Drosophila* develops compound eyes from the eye-primordium of the instar. The developmental mode of eye formation is completely different. On the other hand, genetic pathways that regulate eye formation in flies and vertebrates are well conserved, and *Pax6* plays a central role in both taxa. The mouse mutation *small-eye* and the fly mutation *eye-less* are caused by *Pax6* deficiency. In experiments to examine functional conservation of *Pax6*, cnidarian, protostome, or chordate *Pax6* is introduced into the fly by a sophisticated technique. As in the case of ectopic expression of the fly *Pax6*, the artificially introduced chordate gene resulted in the production of ectopic eyes in the fly. However, the resultant eyes are always *Drosophila* compound eyes, never vertebrate camera or lens eyes. That is, although the function of the master gene is conserved, the downstream structural genes that are engaged in practical formation of the eye differ between flies and vertebrates.

Similar results have been obtained in experiments to confirm conserved roles of transcription factor genes in various processes of organ formation. That is, extant metazoans each have a history of approximately 520 million years for the conserved combinatorial roles of transcription factors/signal pathway molecules and their downstream structural genes. Therefore without determining the relationships or GRNs of toolkit genes and downstream structural genes, we will not be able to achieve a real understanding of the evolution of metazoan body plans.

11.7 THE PHYLOTYPIC STAGE

Modification of preexisting structures and the appearance of novel structures in metazoan bodies become apparent during embryogenesis (eg, Gould, 1977). The concept of the phylotypic stage is interesting and important in discussing the relationship of embryonic development and phylogeny globally. The concept was first suggested by Karl von Baer (1828) and later more clearly schematized by Denis Duboule (1994). Namely, when embryogenesis is compared among five vertebrate taxa (fish, amphibians, reptiles, birds, and mammals), it is evident that eggs and early embryos, as well as adult morphology, are different. Nevertheless, all vertebrates show similar morphology at one specific stage during embryogenesis. This is called the "phylotypic stage" and suggests that a strong constraint maintains it in all vertebrates.

Kalinka and Tomancak (2012) proposed four models explaining patterns of embryonic conservation (Fig. 11.6):

1. *Early conservation (Fig. 11.6A)*: In this model, the earliest developmental stages are considered foundational, and any apparent conservation in later stages is the delayed realization of conservation of genes and proteins acting early (Richardson et al., 1997, 1998; Comte et al., 2010).
2. *Hourglass model (Fig. 11.6B)*: Conservation is greatest in mid-embryogenesis and is either a result of the need for coordination between growth and patterning when the body plan is being built (Duboule, 1994) or the result of a global increase in the complexity of interactions between genes and processes during the phylotypic period (Raff, 1996).
3. *Adaptive penetrance (Fig. 11.6C)*: This model posits that the most important beneficial mutations are likely to occur during the phylotypic period, precisely because this is when the body plan is established (Richardson et al., 1997).
4. *Ontogenetic adjacency (Fig. 11.6D)*: This model suggests that small changes are most likely to occur between adjacent events in the developmental sequence (Poe and Wake, 2004).

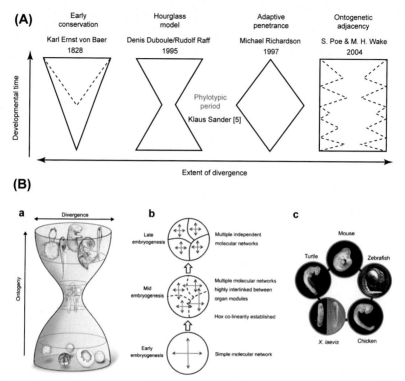

FIGURE 11.6 **Schematic view of the developmental hourglass model.** (A) A schematic comparing four different models that posit different patterns of conservation during animal embryogenesis. In all of these models, development from egg to adult is shown on the y-axis, and evolutionary divergence is represented on the x-axis. *(Modified from Kalinka, A.T., Tomancak, P., 2012. The evolution of early animal embryos: conservation or divergence? Trends in Ecology and Evolution 27, 385–393.)* (B) The developmental hourglass model predicts that mid-embryonic organogenetic stages (the phylotypic period) represent the stage of highest conservation, and that the phylotypic period is the source of the basic body plan at the phylum level. *(From Irie, N., Kuratani, S., 2011. Comparative transcriptome analysis reveals vertebrate phylotypic period during organogenesis. Nature Communications 2, 248.)* a, Hourglass-like divergence has been proposed to result from the spatiotemporal colinearity of Hox cluster genes (Duboule, 1994), and b, from the existence of highly interdependent molecular networks at the phylotypic stage (Raff, 1996). c, Potential phylotypic period for vertebrates. Two stages of *X. laevis* are shown because there was no statistically significant difference between these two stages. *(Modified from Irie, N., Kuratani, S., 2014. The developmental hourglass model: a predictor of the basic body plan? Development 141, 4649–4655.)*

The existence of the phylotypic stage has recently been challenged by genetic analyses (Domazet-Lošo and Tautz, 2010; Irie and Kuratani, 2014). For example, Irie and Kuratani (2011) and Wang et al. (2013) performed comparative transcriptomic analyses of several embryonic stages of zebrafish, *Xenopus*, turtles, chicks, and mice and found that the pharyngula stage is the phylotypic stage of vertebrates. The pharyngula stage is characterized by formation of a head, pharyngeal arches, somites, a neural tube, epidermis, kidney tubules and longitudinal kidney ducts (but no metanephros), a heart with chambers, at least a transient cloaca, no middle ear, no gills on pharyngeal segments, no tongue, no penis or uterus, and no hair or feathers. In addition, approximately 110 genes show conserved expression during the phylotypic stage. These include Hox genes, those involved in cell–cell signaling and interactions (eg, *Fzd2, ptch2, Sema3d, FGFRL1,* and *Dscaml*), transcription factors (eg, *FoxG1, Pax6, myf6, Tbox20, Islet1, Emx2,* and *Klf2*), and secreted morphogens or growth factors (eg, *Dkk, FGF8, Angpt1,* and *INS*). Although the important question of how the expression of these genes is regulated globally has not been answered yet, the developmental hourglass model has been supported not only in vertebrates, but also in *Drosophila* (Kalinka et al., 2010) and *Caenorhabditis* (Levin et al., 2012). Further support comes from plants (Quint et al., 2012) and fungi (Cheng et al., 2015).

Vertebrates evolved dynamically and drastically from a common gnathostome ancestor during the Silurian Period and later. It is understandable that during evolution, strong constraints prevented modification of the body plan during the phylotypic stage because the body plan is essential for development from different egg types to form different adult body plans. An intriguing question is whether it is possible to extrapolate this concept to chordates, ambulacrarians, or deuterostomes. If such stages exist, then what stages in invertebrate chordates, ambulacrarians, or deuterostomes correspond to the vertebrate phylotypic stage? If it is a pharynx-formation stage, then this stage occurs after metamorphosis.

11.8 CONCLUSIONS

The evolution of chordates has been viewed macroscopically. A restructuring of chordate phylogeny is proposed for future investigations of chordate evolution. The phylotypic stage is discussed in relation to deuterostome evolution. In addition, a broader view of comparative genomics and evolutionary developmental biology was discussed in relation to the expression and function of (1) genes acquired by horizontal transfer in the tunicate and/or deuterostome stem; (2) evolution of new genes, perhaps by gene duplication; and (3) domain shuffling of proteins. I also emphasize the significance of downstream structural genes that are involved in the practical establishment of organs and tissues. There are many examples of each, and future exploration of molecular mechanisms involved in expression and function of these genes may give us more insight into the origins and evolution of chordates under the concept of the new organizers hypothesis.

Chapter 12

Summary and Perspective

12.1 SUMMARY

This book began with a general introduction of the five deuterostome groups that were associated with the origin of chordates (Chapter 1). Those include echinoderms, hemichordates, cephalochordates, tunicates (urochordates), and vertebrates. The first two are nonchordate deuterostomes, which together are known as the Ambulacraria, whereas the latter three are chordate deuterostomes. There are discrete differences between ambulacrarians and chordates with regard to their modes of embryogenesis, their larval morphology, and their adult anatomy. Of the three chordate taxa, cephalochordates likely represent something close to the ancestral chordate whereas hemichordates, especially acorn worms, exhibit characters that superficially resemble those of chordates. On the other hand, echinoderms and tunicates are unique in that they have developed highly specialized features, an exoskeleton, and a tunic, respectively. With the aid of new structures, each appears to have evolved independently and dynamically. Vertebrates are special in terms of metazoan evolution. They evolved a head, jaws, and an endoskeleton, although some of these features are not present in the most basal vertebrates, including cyclostomes. The origin and evolution of vertebrates itself is a subject that has been discussed in many articles and books.

Chordate origins are deeply involved in or directly associated with the evolution of vertebrates, including man. Since publication of *The Origin of Species by Means of Natural Selection* by Charles Darwin in 1859, the problem of chordate evolution has been extensively investigated and debated. There have been so many hypotheses to explain chordate origins. At one time or another, almost all bilaterian groups have been suggested as ancestral to chordates. In Chapter 2, representative hypotheses were introduced, including the annelid hypothesis, the auricularia hypothesis, the enteropneust hypothesis, the calcichordate hypothesis, the dorsoventral (DV)-axis inversion hypotheses, and the aboral-dorsalization hypothesis. In this book, the latter proposal has been revised as the new organizers hypothesis. The first three were proposed nearly 100 years ago whereas the last three are relatively recent. Each has its merits—some based on comparisons of adult morphology, some on comparisons of embryos and larvae, and some on comparisons of larval and adult morphologies. Anatomy, embryology, and paleontology have been the main sources of data for inferring classic evolutionary scenarios.

Chordate Origins and Evolution. http://dx.doi.org/10.1016/B978-0-12-802996-1.00012-2

Then, to consider these scenarios more in greater detail, the subsequent three chapters assessed recent progress in the fields of paleontology (Chapter 3), molecular phylogeny (Chapter 4), and comparative genomics (Chapter 5). Recent discoveries have resulted in new interpretations of soft-body fossils that are thought to belong to deuterostomes. Those of enteropneust hemichordates particularly appear to fill a paleontological gap among deuterostome taxa. Current paleontology suggests that all five groups of deuterostomes have their origins in the Early Cambrian Period. In addition, combined paleontology and molecular clock data indicate that all deuterostomes emerged on the earth approximately 570 million years ago or earlier. Another important suggestion is that cephalochordates emerged earlier or at least at almost the same time as hemichordates and echinoderms. Molecular phylogeny based on not only gene and protein sequence comparisons, but also comparisons of other data, including genomic synteny, indels, etc., have provided clear answers. First, the five deuterostome taxa are monophyletic. Second, ambulacrarians and chordates are also each monophyletic. Third, cephalochordates were the first chordate line to diverge; hence, lancelets retain what may be the most primitive characters among chordates. The robust monophyly of deuterostomes indicates that there is no further need to consider protostomes relative to chordate origins.

Genomes of representative species of all five deuterostome taxa have been decoded. Comparative genomics following decoding of the hemichordate genome has succeeded in identifying a gene cluster consisting of *Nk2.1*, *Nk2.2*, *Pax1/9*, and *FoxA* that is deeply involved in the development of pharynx-related structures in all deuterostomes. Although several features, including the mode of early embryogenesis, were once proposed to distinguish deuterostomes from protostomes, a division supported by molecular phylogeny, no definitive structures with genetic backgrounds had previously been identified. Now it is evident that deuterostomes are metazoans with pharyngeal gill slits, the development of which is supported by the pharyngeal gene cluster. Comparative genomics also tells us there is a high grade of synteny conservation between cephalochordates and vertebrates, providing strong evidence for common genetic tools shared by the two lineages. On the other hand, the genomic constitution of vertebrates is unique compared with that of invertebrates, and two rounds of genome-wide gene duplication suggest that the complexity of the vertebrate body plan was accomplished by expansion of the genome. Comparative genomics also shows that tunicates are an offshoot of the main chordate lineage. As filter-feeding specialists, tunicate genomes have been reduced, especially in larvaceans. Nevertheless, in contrast to cephalochordates, which may have retained a body plan close to the ancestral form, tunicates have evolved more complex traits, closer to those of vertebrates than of cephalochordates. Genomics also suggests a significant role of horizontal gene transfer, which has been underestimated in metazoan evolution.

Taking advantage of recent advances in various research fields, Chapter 6 reexamined the validity of hypotheses introduced in Chapter 2 in light of current knowledge. Although each hypothesis includes legitimate points pertinent to the evolutionary history of chordates, it is evident that recent genetic and molecular data do not support the annelid hypothesis, the auricularia hypothesis, or the calcichordate hypothesis. On the other hand, the remaining two hypotheses—the enteropneust hypothesis and the inversion hypotheses—contain several concepts that should be more carefully examined. Therefore, Chapters 7–9 more carefully examined the new organizers hypothesis that I proposed in relation to the enteropneust and inversion hypotheses. First, I reconsidered features that have been thought to characterize chordates: the notochord, the dorsal neural tube, somites, the postanal tail, pharyngeal gill slits, and the endostyle. As discussed in Chapter 6, pharyngeal gill slits are a synapomorphy of deuterostomes because this structure is present in hemichordates and extinct echinoderms. The endostyle is an organ associated with the pharynx to more efficiently capture food. In contrast, the remaining four, especially the notochord, somites, and the postanal tail (but also the dorsal neural tube), are associated with larval locomotion. That is, in contrast to ambulacrarian larvae, which use ciliary locomotion, chordate larvae swim using striated muscles and the notochord to move the tail in an undulatory fashion. In other words, the modification of larval locomotion (the occurrence of fish-like larvae) is the key developmental event in chordate origin.

Further understanding of the origins and evolution of chordates requires basic knowledge of the molecular mechanisms involved in the formation of metazoan body plans. For the last decade, molecular developmental biology has greatly advanced using model organisms, such as *Drosophila* and *Xenopus*. It has become clear that the expression and function of genes that encode transcription factors and signal pathway molecules (genetic toolkit genes) are pivotal in the establishment of metazoan body axes and formation of organs and tissues as well. Chordates share toolkit genes with nonchordate invertebrates. The DV-axis inversion theory is supported by shared, but reversed, Bmp-Chordin (Nodal) gradients between protostomes and deuterostomes. Recent studies reveal that the inversion occurred at the stem of chordates rather than that of deuterostomes. This raises an important question regarding the inversion hypothesis. That is, DV-axis inversion itself does not explain the occurrence of the notochord. Some researchers insist that a homologous organ is present in hemichordates or even in annelids. However, this suggestion conflicts with the timing of notochord formation during early embryogenesis. I previously proposed the aboral-dorsalization hypothesis in which I emphasized this point. However, the aboral-dorsalization hypothesis also fails to explain the occurrence of some chordate characters. This book revises the aboral-dorsalization hypothesis as "the new organizers hypothesis," in which the occurrence of the new organizers in early embryos eventually leads to basic chordate-specific characters. The organizer that leads to the formation of notochord and somites did not occur in ambulacrarian embryos, including the direct-developing enteropneust, *Saccoglossus kowalevskii*. Therefore, formation of new organizers is a key chordate evolutionary event.

On the other hand, the enteropneust hypothesis, irrespective of the incorporation of DV-axis inversion, raises another intriguing question. Why do enteropneust hemichordates resemble chordates or higher vertebrates so closely? Not only anatomy but also a suite of toolkit genes involved in organ formation appears to be shared by hemichordates and chordates. As molecular phylogeny and embryology suggest, hemichordates are ambulacrarians, completely separated from chordates. Then why do they look so similar? Chapter 9 discussed this issue of evolutionary developmental biology.

Chordates appear to have selected three independent evolutionary routes (Chapter 10). Extant cephalochordate lancelets seem to have kept their original (ancestral) form, as seen in the fossil record, suggesting a good match of body plan and lifestyle. On the other hand, urochordates and vertebrates appear to have changed much more dynamically to adapt to different lifestyles, the former for advanced filter feeding and the latter as predators. The evolutionary diversification of tunicates includes the development of a precocious mode of embryogenesis, asexual reproduction, colony formation, etc. The evolutionary diversification of vertebrate anatomy and physiology is so huge that it would require multiple book chapters. Finally, in Chapter 11 I discussed chordate evolution from a broader point of view. I proposed that cephalochordates, urochordates, and vertebrates should not be treated as three subphyla of the phylum Chordata, as in the traditional classification. Because these taxa have characteristic features that distinguish them, the three chordate groups should be classified as phyla. The current classification of deuterostomes as Echinoderms, Hemichordates, and Chordates is somewhat misleading. More natural systems would be Ambulacrarians and Chordates, or Echinoderms, Hemichordates, Cephalochordates, Urochordates, and Vertebrates. Several genetic and molecular mechanisms involved in modifications associated with deuterostome evolution have also been discussed.

In summary, I described in this book everything that I wish to say on the origins and evolution of chordates. Many sections rely on earlier work by other researchers, for which I am deeply indebted. Some sections are still conceptually rough and will require further revision in the future. However, the volume suffices to propound the new organizers hypothesis of chordate origins.

12.2 PERSPECTIVE

Of the five deuterostome taxa, echinoderms and hemichordates are nonchordates whereas cephalochordates, urochordates, and vertebrates are chordates. As discussed earlier, echinoderms and urochordates appear to have evolved independently by developing an exoskeleton and a tunic, respectively. The remaining three taxa are likely the real key to answer the question of chordate origins, which may lie in comparisons of hemichordates and cephalochordates, whereas questions regarding vertebrate evolution may be settled by comparisons of cephalochordates and vertebrates.

However, some caveats are in order. If results suggested by the paleontology-based molecular clock are correct, that is that cephalochordates diverged from the deuterostome ancestor earlier than ambulacrarians, then the first comparison may be less meaningful, although if hemichordates branched earlier (or at the same time) than cephalochordates, then it may prove useful after all. However, because we have no information on the real cephalochordate ancestor (the ancestor before the appearance of related fossils) and because hemichordates and cephalochordates show similar patterns of very early embryogenesis, including egg cleavages, comparison of the two taxa is important to further understand chordate origins. The developmental stage to be compared is also important. If the occurrence of the new organizers is critical for chordate origins, then the comparison of embryos is essential because the organizers arise during late blastula and early gastrula stages. Recent studies of enteropneust hemichordates have mostly focused on anatomy and toolkit gene expression profiles in adults, not embryos. Adult studies do not always provide answers about chordate origins, although we have obtained a great quantity of useful data from them. Much more attention should be focused on comparisons of early embryogenesis in the two taxa. On the other hand, comparisons of hemichordate and cephalochordate adult morphology suggest another intriguing evolutionary question. That is, how can two unrelated taxa with different modes of embryogenesis develop adults that are extremely similar? Simply saying that this is a result of convergent evolution is not adequate. Answering this question will require investigation of many research subjects.

On the other hand, as discussed in Chapter 10, keys to the origin of vertebrates are found in comparisons of cephalochordates and vertebrates. The new leading theory points out the significance of novel developmental systems, including the complexity of the central nervous system, the neural crest, placodes, the cardiopharyngeal field, etc. In contrast to other evolutionary processes discussed in the book, the evolution of vertebrates from boneless groups to bony groups likely took more time. We need to identify the molecular mechanisms involved in these processes by one by one.

The presence of pharyngeal gill slits at the deuterostome stem suggests that the deuterostome ancestor was a filter feeder. The appearance of endostyle-like food-trapping organs in basal chordates and primitive vertebrates implies that chordates adopted a strategy to become more efficient filter feeders. Of several possible strategies to accomplish this aim, the chordate ancestor succeeded in modifying the mode of embryogenesis, from dipleurula-type larvae to fish-like larvae. The new organizers were essential to eventually form fish-like larvae with the notochord, somites, and dorsal hollow neural tube. Having succeeded with this strategy, cephalochordates have retained their ancestral form for more than 520 million years.

On the other hand, one chordate lineage acquired horizontally transferred bacterial genes to make cellulose. With this capacity, the urochordate ancestor produced the tunic, which may function as body armor. Taking advantage of this, the urochordate ancestor abandoned swimming and developed papil-lae at the anterior end, which the larvae use to attach to substrates when they metamorphose to sessile adults. They became proficient filter feeders, trapping food from seawater that enters the oral siphon, passes through pharyngeal gills with the endostyle, and exits via the atrial siphon. This sedentary lifestyle does not consume much energy compared with swimming. When tunicates became sessile, they also changed every developmental process to reach the adult stage as early as possible. To utilize appropriate substrates rapidly after hatching, urochordate tadpole-type larvae never open their mouths before metamorpho-sis. Their developmental mode is mosaic and differentiation of larval muscle is dependent on a maternally provided factor or muscle determinant. I am an ascid-ian developmental biologist, and it was very surprising when we discovered that genes encoding muscle structural proteins (actins and the myosin heavy chain) commence expression at the 64-cell stage, far sooner than actual muscle forma-tion. If every developmental process of tunicates occurs precociously, then this would be understandable.

In contrast, another chordate lineage selected evolved active predation. To use this lifestyle, chordates required modification of the lancelet-like ancestral form to a more active fish-like form. They needed a more complicated suite of organs and tissues, vertebrate-specific innovations, such as a neural crest and placodes, to form a new head and jaws to satisfy these requirements.

Needless to say, chordate evolution has been accomplished with a combina-tion of many developmental and physiological processes. At present, it is highly likely that development of the new organizers was deeply involved in modifica-tion of early embryogenesis and promoted formation of a fish-like body plan with the notochord, somites, and the dorsal hollow neural tube. However, so many questions on the origins of chordates remain open. For example, this book did not address associated changes in physiology, endocrinology, or behavior biology. The really challenging field of integrative biology is just beginning to open.

References

Abitua, P.B., Gainous, T.B., Kaczmarczyk, A.N., Winchell, C.J., Hudson, C., Kamata, K., Nakagawa, M., Tsuda, M., Kusakabe, T.G., Levine, M., 2015. The pre-vertebrate origins of neurogenic placodes. Nature 524, 462–465.

Abitua, P.B., Wagner, E., Navarrete, I.A., Levine, M., 2012. Identification of a rudimentary neural crest in a non-vertebrate chordate. Nature 492, 104–107.

Agius, E., Oelgeschläger, M., Wessely, O., Kemp, C., De Robertis, E.M., 2000. Endodermal Nodal-related signals and mesoderm induction in *Xenopus*. Development 127, 1173–1183.

Aguinaldo, A.M., Turbeville, J.M., Linford, L.S., Rivera, M.C., Garey, J.R., Raff, R.A., Lake, J.A., 1997. Evidence for a clade of nematodes, arthropods and other moulting animals. Nature 387, 489–493.

Ahlberg, P.E., 2001. Major Events in Early Vertebrate Evolution : Palaeontology, Phylogeny, Genetics, and Development. Taylor & Francis, London, UK.

Ahn, D., Ho, R.K., 2008. Tri-phasic expression of posterior *Hox* genes during development of pectoral fins in zebrafish: implications for the evolution of vertebrate paired appendages. Developmental Biology 322, 220–233.

Alföldi, J., Di Palma, F., Grabherr, M., Williams, C., Kong, L., Mauceli, E., Russell, P., Lowe, C.B., Glor, R.E., Jaffe, J.D., Ray, D.A., Boissinot, S., Shedlock, A.M., Botka, C., Castoe, T.A., Colbourne, J.K., Fujita, M.K., Moreno, R.G., ten Hallers, B.F., Haussler, D., Heger, A., Heiman, D., Janes, D.E., Johnson, J., de Jong, P.J., Koriabine, M.Y., Lara, M., Novick, P.A., Organ, C.L., Peach, S.E., Poe, S., Pollock, D.D., de Queiroz, K., Sanger, T., Searle, S., Smith, J.D., Smith, Z., Swofford, R., Turner-Maier, J., Wade, J., Young, S., Zadissa, A., Edwards, S.V., Glenn, T.C., Schneider, C.J., Losos, J.B., Lander, E.S., Breen, M., Ponting, C.P., Lindblad-Toh, K., 2011. The genome of the green anole lizard and a comparative analysis with birds and mammals. Nature 477, 587–591.

Amemiya, C.T., Prohaska, S.J., Hill-Force, A., Cook, A., Wasserscheid, J., Ferrier, D.E., Pascual-Anaya, J., Garcia-Fernàndez, J., Dewar, K., Stadler, P.F., 2008. The amphioxus *Hox* cluster: characterization, comparative genomics, and evolution. Journal of Experimental Zoology Part B: Molecular and Developmental Evolution 310, 465–477.

Angerer, L.M., Yaguchi, S., Angerer, R.C., Burke, R.D., 2011. The evolution of nervous system patterning: insights from sea urchin development. Development 138, 3613–3623.

Annona, G., Holland, N.D., D'Aniello, S., 2015. Evolution of the notochord. EvoDevo 6, 30.

Annunziata, R., Martinez, P., Arnone, M.I., 2013. Intact cluster and chordate-like expression of ParaHox genes in a sea star. BMC Biology 11, 68.

Aparicio, S., Chapman, J., Stupka, E., Putnam, N., Chia, J.M., Dehal, P., Christoffels, A., Rash, S., Hoon, S., Smit, A., Gelpke, M.D., Roach, J., Oh, T., Ho, I.Y., Wong, M., Detter, C., Verhoef, F., Predki, P., Tay, A., Lucas, S., Richardson, P., Smith, S.F., Clark, M.S., Edwards, Y.J., Doggett, N., Zharkikh, A., Tavtigian, S.V., Pruss, D., Barnstead, M., Evans, C., Baden, H., Powell, J., Glusman, G., Rowen, L., Hood, L., Tan, Y.H., Elgar, G., Hawkins, T., Venkatesh, B., Rokhsar, D., Brenner, S., 2002. Whole-genome shotgun assembly and analysis of the genome of *Fugu rubripes*. Science 297, 1301–1310.

Arendt, D., Nübler-Jung, K., 1994. Inversion of dorsoventral axis? Nature 371, 26.

Arendt, D., Nübler-Jung, K., 1999. Comparison of early nerve cord development in insects and vertebrates. Development 126, 2309–2325.

Aszódi, A., Chan, D., Hunziker, E., Bateman, J.F., Fässler, R., 1998. Collagen II is essential for the removal of the notochord and the formation of intervertebral discs. Journal of Cell Biology 143, 1399–1412.

Azumi, K., De Santis, R., De Tomaso, A., Rigoutsos, I., Yoshizaki, F., Pinto, M.R., Marino, R., Shida, K., Ikeda, M., Ikeda, M., Arai, M., Inoue, Y., Shimizu, T., Satoh, N., Rokhsar, D.S., Du Pasquier, L., Kasahara, M., Satake, M., Nonaka, M., 2003. Genomic analysis of immunity in a Urochordate and the emergence of the vertebrate immune system: "waiting for Godot". Immunogenetics 55, 570–581.

Baker, C.V., 2008. The evolution and elaboration of vertebrate neural crest cells. Current Opinion in Genetics and Development 18, 536–543.

Balfour, F.M., 1880–1881. A Treatise on Comparative Embryology. Two Volumes. Macmillan, London.

Balser, E.J., Ruppert, E.E., 1990. Structure, ultrastructure, and function of the preoral heart-kidney in *Saccoglossus kowalevskii* (Hemichordata, Enteropneusta) including new data on the stomochord. Acta Zoologica 71, 235–249.

Bardack, D., 1991. First fossil hagfish (Myxinoidea): a record from the Pennsylvanian of Illinois. Science 254, 701–703.

Bardack, D., Zangerl, R., 1968. First fossil lamprey: a record from the Pennsylvanian of Illinois. Science 162, 1265–1267.

Bateson, W., 1886. The ancestry of the Chordata. Quarterly Journal of Microscopical Science 26, 535–571.

Baughman, K.W., McDougall, C., Cummins, S.F., Hall, M., Degnan, B.M., Satoh, N., Shoguchi, E., 2014. Genomic organization of Hox and ParaHox clusters in the echinoderm, *Acanthaster planci*. Genesis 52, 952–958.

Beaster-Jones, L., Kaltenbach, S.L., Koop, D., Yuan, S., Chastain, R., Holland, L.Z., 2008. Expression of somite segmentation genes in amphioxus: a clock without a wavefront? Development Genes and Evolution 218, 599–611.

Beck, C.W., 2015. Development of the vertebrate tailbud. Wiley Interdisciplinary Reviews: Developmental Biology 4, 33–44.

Bengtson, S., 2005. Mineralized skeletons and early animal evolution. In: Briggs, D.E.G. (Ed.), Evolving Form and Function: Fossils and Development, Peabody Museum of Natural History. Yale University, New Haven, pp. 101–124.

Benton, M.J., 2005. Vertebrate Palaeontology, third ed. Blackwell Science, Oxford.

Berrill, N.J., 1955. The Origin of Vertebrates. Oxford University Press, Oxford, UK.

Bertrand, S., Escriva, H., 2011. Evolutionary crossroads in developmental biology: amphioxus. Development 138, 4819–4830.

Birenheide, R., Tamori, M., Motokawa, T., Ohtani, M., Iwakoshi, E., Muneoka, Y., Fujita, T., Minakata, H., Nomoto, K., 1998. Peptides controlling stiffness of connective tissue in sea cucumbers. Biological Bulletin 194, 253–259.

Bishop, C.D., Burke, R.D., 2007. Ontogeny of the holothurian larval nervous system: evolution of larval forms. Development Genes and Evolution 217, 585–592.

Blair, J.E., Hedges, S.B., 2005. Molecular phylogeny and divergence times of deuterostome animals. Molecular Biology and Evolution 22, 2275–2284.

Blum, M., Feistel, K., Thumberger, T., Schweickert, A., 2014. The evolution and conservation of left-right patterning mechanisms. Development 141, 1603–1613.

Boehm, T., 2012. Evolution of vertebrate immunity. Current Biology 22, R722–R732.

Bone, Q., 1960. The origin of the chordates. Journal of the Linnean Society of London. Zoology 44, 252–269.

Boorman, C.J., Shimeld, S.M., 2002. Pitx homeobox genes in *Ciona* and amphioxus show left-right asymmetry is a conserved chordate character and define the ascidian adenohypophysis. Evolution and Development 4, 354–365.

Boto, L., 2014. Horizontal gene transfer in the acquisition of novel traits by metazoans. Proceedings of the Royal Society B: Biological Sciences 281, 20132450.

Bottjer, D.J., Davidson, E.H., Peterson, K.J., Cameron, R.A., 2006. Paleogenomics of echinoderms. Science 314, 956–960.

Bourlat, S.J., Juliusdottir, T., Lowe, C.J., Freeman, R., Aronowicz, J., Kirschner, M., Lander, E.S., Thorndyke, M., Nakano, H., Kohn, A.B., Heyland, A., Moroz, L.L., Copley, R.R., Telford, M.J., 2006. Deuterostome phylogeny reveals monophyletic chordates and the new phylum Xenoturbellida. Nature 444, 85–88.

Bourlat, S.J., Nielsen, C., Lockyer, A.E., Littlewood, D.T., Telford, M.J., 2003. *Xenoturbella* is a deuterostome that eats molluscs. Nature 424, 925–928.

Brazeau, M.D., Friedman, M., 2015. The origin and early phylogenetic history of jawed vertebrates. Nature 520, 490–497.

Brennan, J., Norris, D.P., Robertson, E.J., 2002. Nodal activity in the node governs left-right asymmetry. Genes and Development 16, 2339–2344.

Bronner, M.E., LeDouarin, N.M., 2012. Development and evolution of the neural crest: an overview. Developmental Biology 366, 2–9.

Brooke, N.M., Garcia-Fernàndez, J., Holland, P.W.H., 1998. The ParaHox gene cluster is an evolutionary sister of the Hox gene cluster. Nature 392, 920–922.

Brown, F.D., Prendergast, A., Swalla, B.J., 2008. Man is but a worm: chordate origins. Genesis 46, 605–613.

Brown, T.A., 2002. Genomes, second ed. Wiley-Liss, Oxford.

Brozovic, M., Martin, C., Dantec, C., Dauga, D., Mendez, M., Simion, P., Percher, M., Laporte, B., Scornavacca, C., Di Gregorio, A., et al., 2016. ANISEED 2015: a digital framework for the comparative developmental biology of ascidians. Nucleic Acids Research 44, D808–818.

Brusca, R.C., Brusca, G.J., 2003. Invertebrates, second ed. (Sunderland).

Burke, A.C., Nelson, C.E., Morgan, B.A., Tabin, C., 1995. *Hox* genes and the evolution of vertebrate axial morphology. Development 121, 333–346.

Cameron, C.B., 2005. A phylogeny of the hemichordates based on morphological characters. Canadian Journal of Zoology 83, 196–215.

Cameron, C.B., Garey, J.R., Swalla, B.J., 2000. Evolution of the chordate body plan: new insights from phylogenetic analyses of deuterostome phyla. Proceedings of the National Academy of Sciences of the United States of America 97, 4469–4474.

Cameron, R.A., Kudtarkar, P., Gordon, S.M., Worley, K.C., Gibbs, R.A., 2015. Do echinoderm genomes measure up? Marine Genomics 22, 1–9.

Cameron, R.A., Rowen, L., Nesbitt, R., Bloom, S., Rast, J.P., Berney, K., Arenas-Mena, C., Martínez, P., Lucas, S., Richardson, P.M., Davidson, E.H., Peterson, K.J., Hood, L., 2006. Unusual gene order and organization of the sea urchin Hox cluster. Journal of Experimental Zoology Part B: Molecular and Developmental Evolution 306b, 45–58.

Candiani, S., Pennati, R., Oliveri, D., Locascio, A., Branno, M., Castagnola, P., Pestarino, M., De Bernardi, F., 2005. *Ci-POU-IV* expression identifies PNS neurons in embryos and larvae of the ascidian *Ciona intestinalis*. Development Genes and Evolution 215, 41–45.

Cañestro, C., Bassham, S., Postlethwait, J.H., 2008. Evolution of the thyroid: anterior-posterior regionalization of the *Oikopleura* endostyle revealed by *Otx*, *Pax2/5/8*, and *Hox1* expression. Developmental Dynamics 237, 1490–1499.

Cannon, J.T., Rychel, A.L., Eccleston, H., Halanych, K.M., Swalla, B.J., 2009. Molecular phylogeny of hemichordata, with updated status of deep-sea enteropneusts. Molecular Phylogenetics and Evolution 52, 17–24.

Cannon, J.T., Swalla, B.J., Halanych, K.M., 2013. Hemichordate molecular phylogeny reveals a novel cold-water clade of harrimaniid acorn worms. Biological Bulletin 225, 194–204.

Caron, J.B., Morris, S.C., Cameron, C.B., 2013. Tubicolous enteropneusts from the Cambrian period. Nature 495, 503–506.

Carroll, S.B., Grenier, J.K., Weatherbee, S.D., 2005. From DNA to Diversity: Molecular Genetics and the Evolution of Animal Design, second ed. Blackwell, Oxford.

Chen, J.Y., Dzik, J., Edgecombe, G.D., Ramskold, L., Zhou, G.Q., 2002a. A possible early Cambrian chordate. Nature 377, 720–722.

Chen, J.Y., Huang, D.Y., Peng, Q.Q., Chi, H.M., Wang, X.Q., Feng, M., 2003. The first tunicate from the early Cambrian of South China. Proceedings of the National Academy of Sciences of the United States of America 100, 8314–8318.

Chen, J.Y., Oliveri, P., Gao, F., Dornbos, S.Q., Li, C.W., Bottjer, D.J., Davidson, E.H., 2002b. Precambrian animal life: probable developmental and adult cnidarian forms from Southwest China. Developmental Biology 248, 182–196.

Chen, L., Xiao, S., Pang, K., Zhou, C., Yuan, X., 2014. Cell differentiation and germ-soma separation in Ediacaran animal embryo-like fossils. Nature 516, 238–241.

Cheng, X., Hui, J.H., Lee, Y.Y., Wan Law, P.T., Kwan, H.S., 2015. A "developmental hourglass" in fungi. Molecular Biology and Evolution 32, 1556–1566.

Chesley, P., 1935. Development of the short-tailed mutant in the house mouse. Journal of Experimental Zoology 70, 429–459.

Chiba, S., Awazu, S., Itoh, M., Chin-Bow, S.T., Satoh, N., Satou, Y., Hastings, K.E.M., 2003. A genomewide survey of developmentally relevant genes in *Ciona intestinalis* – IX. Genes for muscle structural proteins. Development Genes and Evolution 213, 291–302.

Christiaen, L., Burighel, P., Smith, W.C., Vernier, P., Bourrat, F., Joly, J.S., 2002. *Pitx* genes in Tunicates provide new molecular insight into the evolutionary origin of pituitary. Gene 287, 107–113.

Christiaen, L., Wagner, E., Shi, W., Levine, M., 2009. The sea squirt *Ciona intestinalis*. Cold Spring Harbor Protocols 4.

Clarkson, E.N.K., 1998. Invertebrate Palaeontology and Evolution, fourth ed. Blackwell Science, Oxford.

Comte, A., Roux, J., Robinson-Rechavi, M., 2010. Molecular signaling in zebrafish development and the vertebrate phylotypic period. Evolution and Development 12, 144–156.

Conklin, E.G., 1932. The embryology of amphioxus. Journal of Morphology 54, 69–151.

Conway-Morris, S., 1982. Atlas of the Burgess Shale. Palaeontological Association, London, UK.

Conway-Morris, S., 2003. The Cambrian "explosion" of metazoans and molecular biology: would Darwin be satisfied? International Journal of Developmental Biology 47, 505–515.

Conway-Morris, S., 2006. Darwin's dilemma: the realities of the Cambrian 'explosion'. Philosophical Transactions of the Royal Society B: Biological Sciences 361, 1069–1083.

Conway-Morris, S., Caron, J.B., 2012. *Pikaia gracilens* Walcott, a stem-group chordate from the Middle Cambrian of British Columbia. Biological Reviews of the Cambridge Philosophical Society 87, 480–512.

Conway-Morris, S., Caron, J.B., 2014. A primitive fish from the Cambrian of North America. Nature 512, 419–422.

Cook, C.E., Jiménez, E., Akam, M., Saló, E., 2004. The Hox gene complement of acoel flatworms, a basal bilaterian clade. Evolution and Development 6, 154–163.

Corallo, D., Trapani, V., Bonaldo, P., 2015. The notochord: structure and functions. Cellular and Molecular Life Sciences 72, 2989–3008.

Corbo, J.C., Levine, M., Zeller, R.W., 1997. Characterization of a notochord-specific enhancer from the *Brachyury* promoter region of the ascidian, *Ciona intestinalis*. Development 124, 589–602.

Coutelis, J.B., González-Morales, N., Géminard, C., Noselli, S., 2014. Diversity and convergence in the mechanisms establishing L/R asymmetry in metazoa. EMBO Reports 15, 926–937.

Crow, K.D., Stadler, P.F., Lynch, V.J., Amemiya, C., Wagner, G.P., 2006. The "fish-specific" Hox cluster duplication is coincident with the origin of teleosts. Molecular Biology and Evolution 23, 121–136.

Cunningham, D., Casey, E.S., 2014. Spatiotemporal development of the embryonic nervous system of *Saccoglossus kowalevskii*. Developmental Biology 386, 252–263.

Cuvier, G., 1815. Me´moire sur les Ascidies et sur leur anatomie, 2. Memoirs du Museum d'Histoire Naturelle, Paris, pp. 10–39.

Darwin, C., 1859. On the Origin of Species. John. Murray, London, UK.

Davidson, E.H., 2006. The Regulatory Genome: Gene Regulatory Networks in Development and Evolution. Academic Press, San Diego, CA.

Davidson, E.H., Erwin, D.H., 2006. Gene regulatory networks and the evolution of animal body plans. Science 311, 796–800.

De Robertis, E.M., 2008. Evo-devo: variations on ancestral themes. Cell 132, 185–195.

De Robertis, E.M., Sasai, Y., 1996. A common plan for dorsoventral patterning in Bilateria. Nature 380, 37–40.

Dehal, P., Satou, Y., Campbell, R.K., Chapman, J., Degnan, B., De Tomaso, A., Davidson, B., Di Gregorio, A., Gelpke, M., Goodstein, D.M., Harafuji, N., Hastings, K.E.M., Ho, I., Hotta, K., Huang, W., Kawashima, T., Lemaire, P., Martinez, D., Meinertzhagen, I.A., Necula, S., Nonaka, M., Putnam, N., Rash, S., Saiga, H., Satake, M., Terry, A., Yamada, L., Wang, H.G., Awazu, S., Azumi, K., Boore, J., Branno, M., Chin-bow, S., DeSantis, R., Doyle, S., Francino, P., Keys, D.N., Haga, S., Hayashi, H., Hino, K., Imai, K.S., Inaba, K., Kano, S., Kobayashi, K., Kobayashi, M., Lee, B.I., Makabe, K.W., Manohar, C., Matassi, G., Medina, M., Mochizuki, Y., Mount, S., Morishita, T., Miura, S., Nakayama, A., Nishizaka, S., Nomoto, H., Ohta, F., Oishi, K., Rigoutsos, I., Sano, M., Sasaki, A., Sasakura, Y., Shoguchi, E., Shin-i, T., Spagnuolo, A., Stainier, D., Suzuki, M.M., Tassy, O., Takatori, N., Tokuoka, M., Yagi, K., Yoshizaki, F., Wada, S., Zhang, C., Hyatt, P.D., Larimer, F., Detter, C., Doggett, N., Glavina, T., Hawkins, T., Richardson, P., Lucas, S., Kohara, Y., Levine, M., Satoh, N., Rokhsar, D.S., 2002. The draft genome of *Ciona intestinalis*: insights into chordate and vertebrate origins. Science 298, 2157–2167.

Delaune, E., Lemaire, P., Kodjabachian, L., 2005. Neural induction in *Xenopus* requires early FGF signalling in addition to BMP inhibition. Development 132, 299–310.

Delsuc, F., Brinkmann, H., Chourrout, D., Philippe, H., 2006. Tunicates and not cephalochordates are the closest living relatives of vertebrates. Nature 439, 965–968.

Denes, A.S., Jékely, G., Steinmetz, P.R.H., Raible, F., Snyman, H., Prud'homme, B., Ferrier, D.E.K., Balavoine, G., Arendt, D., 2007. Molecular architecture of annelid nerve cord supports common origin of nervous system centralization in bilateria. Cell 129, 277–288.

Denoeud, F., Henriet, S., Mungpakdee, S., Aury, J.M., Da Silva, C., Brinkmann, H., Mikhaleva, J., Olsen, L.C., Jubin, C., Cañestro, C., Bouquet, J.M., Danks, G., Poulain, J., Campsteijn, C., Adamski, M., Cross, I., Yadetie, F., Muffato, M., Louis, A., Butcher, S., Tsagkogeorga, G., Konrad, A., Singh, S., Jensen, M.F., Huynh Cong, E., Eikeseth-Otteraa, H., Noel, B., Anthouard, V., Porcel, B.M., Kachouri-Lafond, R., Nishino, A., Ugolini, M., Chourrout, P., Nishida, H., Aasland, R., Huzurbazar, S., Westhof, E., Delsuc, F., Lehrach, H., Reinhardt, R., Weissenbach, J., Roy, S.W., Artiguenave, F., Postlethwait, J.H., Manak, J.R., Thompson, E.M., Jaillon, O., Du Pasquier, L., Boudinot, P., Liberles, D.A., Volff, J.N., Philippe, H., Lenhard, B., Roest Crollius, H., Wincker, P., Chourrout, D., 2010. Plasticity of animal genome architecture unmasked by rapid evolution of a pelagic tunicate. Science 330, 1381–1385.

Diogo, R., Kelly, R.G., Christiaen, L., Levine, M., Ziermann, J.M., Molnar, J.L., Noden, D.M., Tzahor, E., 2015. A new heart for a new head in vertebrate cardiopharyngeal evolution. Nature 520, 466–473.

Diogo, R., Ziermann, J.M., 2015. Development, metamorphosis, morphology, and diversity: the evolution of chordate muscles and the origin of vertebrates. Developmental Dynamics 244, 1046–1057.

Di-Poï, N., Montoya-Burgos, J.I., Miller, H., Pourquie, O., Milinkovitch, M.C., Duboule, D., 2010. Changes in *Hox* genes' structure and function during the evolution of the squamate body plan. Nature 464, 99–103.

Dobrovolskaia-Zavadskaia, N., 1927. Sur la mortification spontanée de la queue chez la souris nouveau-née et sur l'existence d'un caractère (facteur) héréditaire "non viable". Comptes Rendus des Seances de la Societe de Biologie tt de ses Filiales 97, 114–116.

Dohrn, A., 1875. Der Ursprung der Wirbelthiere und das Princip des Functionswechsels. Genealogische Skizzen [in German]. Engelmann, Leipzig.

Domazet-Lošo, T., Tautz, D., 2010. A phylogenetically based transcriptome age index mirrors ontogenetic divergence patterns. Nature 468, 815–818.

Duboc, V., Röttinger, E., Besnardeau, L., Lepage, T., 2004. Nodal and BMP2/4 signaling organizes the oral-aboral axis of the sea urchin embryo. Developmental Cell 6, 397–410.

Duboule, D., 1994. Temporal colinearity and the phylotypic progression – a basis for the stability of a vertebrate Bauplan and the evolution of morphologies through heterochrony. Development. Supplement 135–142.

Duboule, D., 2007. The rise and fall of Hox gene clusters. Development 134, 2549–2560.

Dunn, C.W., Giribet, G., Edgecombe, G.D., Hejnol, A., 2014. Animal phylogeny and its evolutionary implications. Annual Review of Ecology, Evolution, and Systematics 45, 371–395.

Dunn, C.W., Hejnol, A., Matus, D.Q., Pang, K., Browne, W.E., Smith, S.A., Seaver, E., Rouse, G.W., Obst, M., Edgecombe, G.D., Sørensen, M.V., Haddock, S.H., Schmidt-Rhaesa, A., Okusu, A., Kristensen, R.M., Wheeler, W.C., Martindale, M.Q., Giribet, G., 2008. Broad phylogenomic sampling improves resolution of the animal tree of life. Nature 452, 745–749.

Dupret, V., Sanchez, S., Goujet, D., Tafforeau, P., Ahlberg, P.E., 2014. A primitive placoderm sheds light on the origin of the jawed vertebrate face. Nature 507, 500–503.

Eakin, R.M., Kuda, A., 1971. Ultrastructure of sensory receptors in Ascidian tadpoles. Zeitschrift für Zellforschung und mikroskopische Anatomie 112, 287–312.

Edlund, R.K., Ohyama, T., Kantarci, H., Riley, B.B., Groves, A.K., 2014. Foxi transcription factors promote pharyngeal arch development by regulating formation of FGF signaling centers. Developmental Biology 390, 1–13.

Erclik, T., Hartenstein, V., McInnes, R.R., Lipshitz, H.D., 2009. Eye evolution at high resolution: the neuron as a unit of homology. Developmental Biology 332, 70–79.

Erwin, D.H., 2011. Evolutionary uniformitarianism. Developmental Biology 357, 27–34.

Erwin, D.H., Laflamme, M., Tweedt, S.M., Sperling, E.A., Pisani, D., Peterson, K.J., 2011. The Cambrian conundrum: early divergence and later ecological success in the early history of animals. Science 334, 1091–1097.

Evans, A.L., Faial, T., Gilchrist, M.J., Down, T., Vallier, L., Pedersen, R.A., Wardle, F.C., Smith, J.C., 2012. Genomic targets of *Brachyury* (*T*) in differentiating mouse embryonic stem cells. PLoS One 7, e33346.

Faure, S., Lee, M.A., Keller, T., ten Dijke, P., Whitman, M., 2000. Endogenous patterns of TGFβ superfamily signaling during early *Xenopus* development. Development 127, 2917–2931.

Ferrier, D.E., Minguillón, C., Holland, P.W., Garcia-Fernàndez, J., 2000. The amphioxus *Hox* cluster: deuterostome posterior flexibility and *Hox14*. Evolution and Development 2, 284–293.

Flot, J.F., Hespeels, B., Li, X., Noel, B., Arkhipova, I., Danchin, E.G.J., Hejnol, A., Henrissat, B., Koszul, R., Aury, J.M., Barbe, V., Barthélémy, R.M., Bast, J., Bazykin, G.A., Chabrol, O., Couloux, A., Da Rocha, M., Da Silva, C., Gladyshev, E., Gouret, P., Hallatschek, O., Hecox-Lea, B., Labadie, K., Lejeune, B., Piskurek, O., Poulain, J., Rodriguez, F., Ryan, J.F., Vakhrusheva, O.A., Wajnberg, E., Wirth, B., Yushenova, I., Kellis, M., Kondrashov, A.S., Welch, D.B.M., Pontarotti, P., Weissenbach, J., Wincker, P., Jaillon, O., Van Doninck, K., 2013. Genomic evidence for ameiotic evolution in the bdelloid rotifer *Adineta vaga*. Nature 500, 453–457.

Freeman, R., Ikuta, T., Wu, M., Koyanagi, R., Kawashima, T., Tagawa, K., Humphreys, T., Fang, G.C., Fujiyama, A., Saiga, H., Lowe, C., Worley, K., Jenkins, J., Schmutz, J., Kirschner, M., Rokhsar, D., Satoh, N., Gerhart, J., 2012. Identical genomic organization of two hemichordate hox clusters. Current Biology 22, 2053–2058.

Fritzsch, G., Böhme, M.U., Thorndyke, M., Nakano, H., Israelsson, O., Stach, T., Schlegel, M., Hankeln, T., Stadler, P.F., 2008. PCR survey of *Xenoturbella bocki* Hox genes. Journal of Experimental Zoology Part B: Molecular and Developmental Evolution 310, 278–284.

Fujiwara, S., Corbo, J.C., Levine, M., 1998. The snail repressor establishes a muscle/notochord boundary in the *Ciona* embryo. Development 125, 2511–2520.

Gans, C., Northcutt, R.G., 1983. Neural crest and the origin of vertebrates: a new head. Science 220, 268–273.

Garcia-Fernàndez, J., 2005a. The genesis and evolution of homeobox gene clusters. Nature Reviews Genetics 6, 881–892.

Garcia-Fernàndez, J., 2005b. Hox, ParaHox, ProtoHox: facts and guesses. Heredity 94, 145–152.

Garstang, M., Ferrier, D.E., 2013. Time is of the essence for ParaHox homeobox gene clustering. BMC Biology 11, 72.

Garstang, W., 1928a. The Origin and Evolution of Larval Forms Report of the British Association for the Advancement of Science No. 1928. , pp. 77–98.

Garstang, W., 1928b. The morphology of the Tunicata, and its bearings on the phylogeny of the Chordata. Quarterly Journal of Microscopical Science 72, 51–187.

Gaskell, W.H., 1890. On the origin of vertebrates from a crustacean-like ancestor. Quarterly Journal of Microscopical Science 31, 379–444.

Gee, H., 1996. Before the Backbone: Views on the Origin of the Vertebrates. Chapman & Hall, London.

Gehring, W.J., 2005. New perspectives on eye development and the evolution of eyes and photoreceptors. Journal of Heredity 96, 171–184.

Gentsch, G.E., Owens, N.D., Martin, S.R., Piccinelli, P., Faial, T., Trotter, M.W., Gilchrist, M.J., Smith, J.C., 2013. In vivo T-box transcription factor profiling reveals joint regulation of embryonic neuromesodermal bipotency. Cell Reports 4, 1185–1196.

Geoffroy St-Hilaire, E., 1822. Considérations générales sur la vertèbre. [in French] Memoirs du Museum d'Histoire Naturelle 9, 89–119.

Gerhart, J., 2000. Inversion of the chordate body axis: are there alternatives? Proceedings of the National Academy of Sciences of the United States of America 97, 4445–4448.

Gerhart, J., 2001. Evolution of the organizer and the chordate body plan. International Journal of Developmental Biology 45, 133–153.

Gerhart, J., 2006. The deuterostome ancestor. Journal of Cellular Physiology 209, 677–685.

Gerhart, J., Lowe, C., Kirschner, M., 2005. Hemichordates and the origin of chordates. Current Opinion in Genetics and Development 15, 461–467.

Gilbert, S.F., 2013. Developmental Biology, tenth ed. Sinauer Associates, Sunderland, MA, USA.

Gillis, J.A., Fritzenwanker, J.H., Lowe, C.J., 2012. A stem-deuterostome origin of the vertebrate pharyngeal transcriptional network. Proceedings of the Royal Society B: Biological Sciences 279, 237–246.

Gomez, C., Ozbudak, E.M., Wunderlich, J., Baumann, D., Lewis, J., Pourquié, O., 2008. Control of segment number in vertebrate embryos. Nature 454, 335–339.

Gont, L.K., Steinbeisser, H., Blumberg, B., de Robertis, E.M., 1993. Tail formation as a continuation of gastrulation: the multiple cell populations of the *Xenopus* tailbud derive from the late blastopore lip. Development 119, 991–1004.

Gorman, A.L., McReynolds, J.S., Barnes, S.N., 1971. Photoreceptors in primitive chordates: fine structure, hyperpolarizing receptor potentials, and evolution. Science 172, 1052–1054.

Gould, S.J., 1977. Ontogeny and Phylogeny. Belknap Press of Harvard University Press, Cambridge.

Graham, A., Richardson, J., 2012. Developmental and evolutionary origins of the pharyngeal apparatus. Evodevo 3, 24.

Graham, L.A., Lougheed, S.C., Ewart, K.V., Davies, P.L., 2008. Lateral transfer of a lectin-like antifreeze protein gene in fishes. PLoS One 3, e2616.

Green, S.A., Norris, R.P., Terasaki, M., Lowe, C.J., 2013. FGF signaling induces mesoderm in the hemichordate *Saccoglossus kowalevskii*. Development 140, 1024–1033.

Green, S.A., Simoes-Costa, M., Bronner, M.E., 2015. Evolution of vertebrates as viewed from the crest. Nature 520, 474–482.

Grobben, K., 1908. Die systematische Einteilung des Tierreiches. Verhandlungen der Zoologisch-Botanischen Gesellschaft in Wien 58, 491–511.

Gross, J.M., McClay, D.R., 2001. The role of *Brachyury* (*T*) during gastrulation movements in the sea urchin *Lytechinus variegatus*. Developmental Biology 239, 132–147.

Guo, P., Hirano, M., Herrin, B.R., Li, J., Yu, C., Sadlonova, A., Cooper, M.D., 2009. Dual nature of the adaptive immune system in lampreys. Nature 459, 796–801.

Hadfield, K.A., Swalla, B.J., Jeffery, W.R., 1995. Multiple origins of anural development in ascidians inferred from rDNA sequences. Journal of Molecular Evolution 40, 413–427.

Haeckel, E., 1866. Generelle Morphologie der Organismen. Verlag von Georg Reimer, Berlin, Germany.

Haeckel, E., 1874a. Anthropogenie, oder, Entwickelungsgeschichte des Menschen. Verlag von Wilhelm Engelmann, Leipzig, Germany.

Haeckel, E., 1874b. Die Gastraea-Theorie, die phylogenetische Classification des Thierreichs und die Homologie der Keimblatter. Jenaische Zischr Naturw 8, 1–55.

Haeckel, E., 1894. Systematische Phylogenie. Verlag von Georg Reimer, Berlin, Germany.

Haillot, E., Molina, M.D., Lapraz, F., Lepage, T., 2015. The Maternal Maverick/GDF15-like TGF-β Ligand Panda Directs Dorsal-Ventral Axis Formation by Restricting Nodal Expression in the Sea Urchin Embryo. PLoS Biology 13, e1002247.

Halanych, K.M., 1995. The phylogenetic position of the pterobranch hemichordates based on 18S rDNA sequence data. Molecular Phylogenetics and Evolution 4, 72–76.

Halanych, K.M., Bacheller, J.D., Aguinaldo, A.M.A., Liva, S.M., Hillis, D.M., Lake, J.A., 1995. Evidence from 18s ribosomal DNA that the lophophorates are protostome animals. Science 267, 1641–1643.

Halanych, K.M., Cannon, J.T., Mahon, A.R., Swalla, B.J., Smith, C.R., 2013. Modern Antarctic acorn worms form tubes. Nature Communications 4, 2738.

Hall, B.K., 1999. Evolutionary Developmental Biology, second ed. Kluwer Academic Publishers, Dordrecht, Netherlands.

Hall, B.K., 2000. The neural crest as a fourth germ layer and vertebrates as quadroblastic not triploblastic. Evolution and Development 2, 3–5.

Hara, Y., Yamaguchi, M., Akasaka, K., Nakano, H., Nonaka, M., Amemiya, S., 2006. Expression patterns of *Hox* genes in larvae of the sea lily *Metacrinus rotundus*. Development Genes and Evolution 216, 797–809.

Hatschek, B., 1893. The Amphioxus and Its Development. Swan Sonnenschein & Co., London.

Heasman, J., 2006. Patterning the early *Xenopus* embryo. Development 133, 1205–1217.

Hedges, S.B., Poling, L.L., 1999. A molecular phylogeny of reptiles. Science 283, 998–1001.

Hejnol, A., Lowe, C.J., 2014. Animal evolution: stiff or squishy notochord origins? Current Biology 24, R1131–R1133.

Hejnol, A., Martindale, M.Q., 2008. Acoel development supports a simple planula-like urbilaterian. Philosophical Transactions of the Royal Society of London. Series B, Biological Sciences 363, 1493–1501.

Hejnol, A., Obst, M., Stamatakis, A., Ott, M., Rouse, G.W., Edgecombe, G.D., Martinez, P., Baguñà, J., Bailly, X., Jondelius, U., Wiens, M., Müller, W.E., Seaver, E., Wheeler, W.C., Martindale, M.Q., Giribet, G., Dunn, C.W., 2009. Assessing the root of bilaterian animals with scalable phylogenomic methods. Proceedings of the Royal Society B: Biological Sciences 276, 4261–4270.

Henry, J.Q., Perry, K.J., Wever, J., Seaver, E., Martindale, M.Q., 2008. β-Catenin is required for the establishment of vegetal embryonic fates in the nemertean, *Cerebratulus lacteus*. Developmental Biology 317, 368–379.

Hinman, V.F., Davidson, E.H., 2007. Evolutionary plasticity of developmental gene regulatory network architecture. Proceedings of the National Academy of Sciences of the United States of America 104, 19404–19409.

Hirakow, R., Kajita, N., 1994. Electron microscopic study of the development of amphioxus, *Branchiostoma belcheri tsingtauense*: the neurula and larva. Acta Anatomica Nipponica 69, 1–13.

Hiruta, J., Mazet, F., Yasui, K., Zhang, P.J., Ogasawara, M., 2005. Comparative expression analysis of transcription factor genes in the endostyle of invertebrate chordates. Developmental Dynamics 233, 1031–1037.

Hoegg, S., Meyer, A., 2005. Hox clusters as models for vertebrate genome evolution. Trends in Genetics 21, 421–424.

Holland, L.Z., 2009. Chordate roots of the vertebrate nervous system: expanding the molecular toolkit. Nature Reviews Neuroscience 10, 736–746.

Holland, L.Z., 2013a. Evolution of new characters after whole genome duplications: insights from amphioxus. Seminars in Cell and Developmental Biology 24, 101–109.

Holland, L.Z., 2015. Evolution of basal deuterostome nervous systems. Journal of Experimental Biology 218, 637–645.

Holland, L.Z., Albalat, R., Azumi, K., Benito-Gutiérrez, E., Blow, M.J., Bronner-Fraser, M., Brunet, F., Butts, T., Candiani, S., Dishaw, L.J., Ferrier, D.E., Garcia-Fernàndez, J., Gibson-Brown, J.J., Gissi, C., Godzik, A., Hallböök, F., Hirose, D., Hosomichi, K., Ikuta, T., Inoko, H., Kasahara, M., Kasamatsu, J., Kawashima, T., Kimura, A., Kobayashi, M., Kozmik, Z., Kubokawa, K., Laudet, V., Litman, G.W., McHardy, A.C., Meulemans, D., Nonaka, M., Olinski, R.P., Pancer, Z., Pennacchio, L.A., Pestarino, M., Rast, J.P., Rigoutsos, I., Robinson-Rechavi, M., Roch, G., Saiga, H., Sasakura, Y., Satake, M., Satou, Y., Schubert, M., Sherwood, N., Shiina, T., Takatori, N., Tello, J., Vopalensky, P., Wada, S., Xu, A., Ye, Y., Yoshida, K., Yoshizaki, F., Yu, J.K., Zhang, Q., Zmasek, C.M., de Jong, P.J., Osoegawa, K., Putnam, N.H., Rokhsar, D.S., Satoh, N., Holland, P.W., 2008. The amphioxus genome illuminates vertebrate origins and cephalochordate biology. Genome Research 18, 1100–1111.

Holland, L.Z., Holland, N.D., 2001. Evolution of neural crest and placodes: amphioxus as a model for the ancestral vertebrate? Journal of Anatomy 199, 85–98.

Holland, L.Z., Holland, N.D., 2007. A revised fate map for amphioxus and the evolution of axial patterning in chordates. Integrative and Comparative Biology 47, 360–372.

Holland, N.D., 2003. Early central nervous system evolution: an era of skin brains? Nature Reviews Neuroscience 4, 617–627.

Holland, N.D., Holland, L.Z., Holland, P.W., 2015. Scenarios for the making of vertebrates. Nature 520, 450–455.

Holland, P.W., 2000. Embryonic development of heads, skeletons and amphioxus: Edwin S. Goodrich revisited. International Journal of Developmental Biology 44, 29–34.

Holland, P.W., 2013b. Evolution of homeobox genes. Wiley Interdisciplinary Reviews. Developmental Biology 2, 31–45.

Holland, P.W., Garcia-Fernàndez, J., Williams, N.A., Sidow, A., 1994. Gene duplications and the origins of vertebrate development. Development Supplement 125–133.

Holland, P.W.H., Koschorz, B., Holland, L.Z., Herrmann, B.G., 1995. Conservation of *Brachyury* (*T*) genes in amphioxus and vertebrates: developmental and evolutionary implications. Development 121, 4283–4291.

Holley, S.A., Jackson, P.D., Sasai, Y., Lu, B., Derobertis, E.M., Hoffmann, F.M., Ferguson, E.L., 1995. A conserved system for dorsal-ventral patterning in insects and vertebrates involving sog and *chordin*. Nature 376, 249–253.

Hotta, K., Takahashi, H., Satoh, N., Gojobori, T., 2008. *Brachyury*-downstream gene sets in a chordate, *Ciona intestinalis*: integrating notochord specification, morphogenesis and chordate evolution. Evolution and Development 10, 37–51.

Hou, X.G., Aldridge, R., Bergstrom, J., Siveter, D.J., Siveter, D., Feng, X.H., 2007. The Cambrian Fossils of Chengjiang, China: The Flowering of Early Animal Life. Wiley-Blackwell.

Huang, S., Chen, Z., Yan, X., Yu, T., Huang, G., Yan, Q., Pontarotti, P.A., Zhao, H., Li, J., Yang, P., Wang, R., Li, R., Tao, X., Deng, T., Wang, Y., Li, G., Zhang, Q., Zhou, S., You, L., Yuan, S., Fu, Y., Wu, F., Dong, M., Chen, S., Xu, A., 2014. Decelerated genome evolution in modern vertebrates revealed by analysis of multiple lancelet genomes. Nature Communications 5, 5896.

Hubaud, A., Pourquié, O., 2014. Signalling dynamics in vertebrate segmentation. Nature Reviews Molecular Cell Biology 15, 709–721.

Hufton, A.L., Mathia, S., Braun, H., Georgi, U., Lehrach, H., Vingron, M., Poustka, A.J., Panopoulou, G., 2009. Deeply conserved chordate noncoding sequences preserve genome synteny but do not drive gene duplicate retention. Genome Research 19, 2036–2051.

Hyman, L.H., 1959. Smaller Coelomate Groups. McGraw-Hill, New York.

Ikuta, T., Chen, Y.C., Annunziata, R., Ting, H.C., Tung, C.H., Koyanagi, R., Tagawa, K., Humphreys, T., Fujiyama, A., Saiga, H., Satoh, N., Yu, J.K., Arnone, M.I., Su, Y.H., 2013. Identification of an intact ParaHox cluster with temporal colinearity but altered spatial colinearity in the hemichordate *Ptychodera flava*. BMC Evolutionary Biology 13, 129.

Ikuta, T., Yoshida, N., Satoh, N., Saiga, H., 2004. *Ciona intestinalis* Hox gene cluster: its dispersed structure and residual colinear expression in development. Proceedings of the National Academy of Sciences of the United States of America 101, 15118–15123.

Imai, K.S., Levine, M., Satoh, N., Satou, Y., 2006. Regulatory blueprint for a chordate embryo. Science 312, 1183–1187.

Irie, N., Kuratani, S., 2011. Comparative transcriptome analysis reveals vertebrate phylotypic period during organogenesis. Nature Communications 2, 248.

Irie, N., Kuratani, S., 2014. The developmental hourglass model: a predictor of the basic body plan? Development 141, 4649–4655.

Jandzik, D., Garnett, A.T., Square, T.A., Cattell, M.V., Yu, J.K., Medeiros, D.M., 2015. Evolution of the new vertebrate head by co-option of an ancient chordate skeletal tissue. Nature 518, 534–537.

Janvier, P., 2015. Facts and fancies about early fossil chordates and vertebrates. Nature 520, 483–489.

Jefferies, R.P.S., 1986. The Ancestry of the Vertebrates. British Museum (Natural History), London, UK.

Jefferies, R.P.S., 1991. In: Bock, G.R., Marsh, J. (Eds.), Biological Asymmetry and Handedness. Wiley, Chichester, pp. 94–127.

Jefferies, R.P.S., Brown, N.A., Daley, P.E.J., 1996. The early phylogeny of chordates and echinoderms and the origin of chordate left-right asymmetry and bilateral symmetry. Acta Zoologica (Stockholm) 77, 101–122.

Jeffery, W.R., Strickler, A.G., Yamamoto, Y., 2004. Migratory neural crest-like cells form body pigmentation in a urochordate embryo. Nature 431, 696–699.

Jeffery, W.R., Swalla, B.J., 1997. Tunicates. In: Gilbert, S.F., Raunio, A.M. (Eds.), Embryology: Constructing the Organism. Sinauer Associates Inc., Mass, USA, pp. 331–364.

Jiang, D., Smith, W.C., 2007. Ascidian notochord morphogenesis. Developmental Dynamics 236, 1748–1757.

Jiménez-Guri, E., Paps, J., García-Fernández, J., Saló, E., 2006. *Hox* and *ParaHox* genes in Nemertodermatida, a basal bilaterian clade. International Journal of Developmental Biology 50, 675–679.

Jollie, M., 1973. The origin of the chordates. Acta Zoologica 54, 81–100.

José-Edwards, D.S., Kerner, P., Kugler, J.E., Deng, W., Jiang, D., Di Gregorio, A., 2011. The identification of transcription factors expressed in the notochord of *Ciona intestinalis* adds new potential players to the brachyury gene regulatory network. Developmental Dynamics 240, 1793–1805.

Kalinka, A.T., Tomancak, P., 2012. The evolution of early animal embryos: conservation or divergence? Trends in Ecology and Evolution 27, 385–393.

Kalinka, A.T., Varga, K.M., Gerrard, D.T., Preibisch, S., Corcoran, D.L., Jarrells, J., Ohler, U., Bergman, C.M., Tomancak, P., 2010. Gene expression divergence recapitulates the developmental hourglass model. Nature 468, 811–814.

Kardong, K.V., 2014. Vertebrates: Comparative Anatomy, Function, Evolution, seventh ed. McGraw-Hill, New York, NY.

Kaul, S., Stach, T., 2010. Ontogeny of the collar cord: neurulation in the hemichordate *Saccoglossus kowalevskii*. Journal of Morphology 271, 1240–1259.

Kawamura, et al., 2015. Histone methylation codes involved in stemness, multipotency, and senescence in budding tunicates. Mechanisms of Ageing and Development 145, 1–12.

Kawashima, T., Kawashima, S., Tanaka, C., Murai, M., Yoneda, M., Putnam, N.H., Rokhsar, D.S., Kanehisa, M., Satoh, N., Wada, H., 2009. Domain shuffling and the evolution of vertebrates. Genome Research 19, 1393–1403.

Kelly, R.G., 2012. The second heart field. Current Topics in Developmental Biology 100, 33–65.

Khalturin, K., Hemmrich, G., Fraune, S., Augustin, R., Bosch, T.C.G., 2009. More than just orphans: are taxonomically-restricted genes important in evolution? Trends in Genetics 25, 404–413.

Kiecker, C., Bates, T., Bell, E., 2015. Molecular specification of germ layers in vertebrate embryos. Cellular and Molecular Life Sciences 73 (5), 923–947.

Kimelman, D., 2006. Mesoderm induction: from caps to chips. Nature Reviews Genetics 7, 360–372.

Kimmel, C.,B., Ballard, W.W., Kimmel, S.R., Ullmann, B., Schilling, T.F., 1995. Stages of embryonic development of the zebrafish. Developmental Dynamics 203, 253–310.

King, B.L., Gillis, J.A., Carlisle, H.R., Dahn, R.D., 2011. A natural deletion of the *HoxC* cluster in elasmobranch fishes. Science 334, 1517.

Kispert, A., Koschorz, B., Herrmann, B.G., 1995. The T-protein encoded by *Brachyury* is a tissue-specific transcription factor. EMBO Journal 14, 4763–4772.

Koide, T., Hayata, T., Cho, K.W.Y., 2005. Xenopus as a model system to study transcriptional regulatory networks. Proceedings of the National Academy of Sciences of the United States of America 102, 4943–4948.

Kon, T., Nohara, M., Yamanoue, Y., Fujiwara, Y., Nishida, M., Nishikawa, T., 2007. Phylogenetic position of a whale-fall lancelet (Cephalochordata) inferred from whole mitochondrial genome sequences. BMC Evolutionary Biology 7, 127.

Kondrashov, F.A., Koonin, E.V., Morgunov, I.G., Finogenova, T.V., Kondrashova, M.N., 2006. Evolution of glyoxylate cycle enzymes in Metazoa: evidence of multiple horizontal transfer events and pseudogene formation. Biology Direct 1, 31.

Kowalevsky, A., 1866a. Entwicklungsgeschichte der einfachen Ascidien. Mémoires de l'Académie impériale des sciences de St. Pétersbourg 7, 11–19.

Kowalevsky, A., 1866b. Anatomie des Balanoglossus. Mémoires de l'Académie Impériale des Sciences de St. Pétersbourg 7, 1–10.

Kowalevsky, A., 1867. Entwicklungsgeschichte des Amphioxus lanceolatus. Mémoires de l'Académie impériale des sciences de St. Pétersbourg 7, 1–17.

Kuroda, H., Fuentealba, L., Ikeda, A., Reversade, B., De Robertis, E.M., 2005. Default neural induction: neuralization of dissociated *Xenopus* cells is mediated by Ras/MAPK activation. Genes and Development 19, 1022–1027.

Kuroda, H., Wessely, O., De Robertis, E.M., 2004. Neural induction in *Xenopus*: requirement for ectodermal and endomesodermal signals via Chordin, Noggin, β-Catenin, and Cerberus. PLoS Biology 2, E92.

Kusakabe, T., Araki, I., Satoh, N., Jeffery, W.R., 1997. Evolution of chordate actin genes: evidence from genomic organization and amino acid sequences. Journal of Molecular Evolution 44, 289–298.

Kusakabe, T., Kusakabe, R., Kawakami, I., Satou, Y., Satoh, N., Tsuda, M., 2001. *Ci-opsin1*, a vertebrate-type opsin gene, expressed in the larval ocellus of the ascidian *Ciona intestinalis*. FEBS Letters 506, 69–72.

Kusakabe, T., Makabe, K.W., Satoh, N., 1992. Tunicate muscle actin genes. Structure and organization as a gene cluster. Journal of Molecular Biology 227, 955–960.

Kusakabe, T., Tsuda, M., 2007. Photoreceptive systems in ascidians. Photochemistry and Photobiology 83, 248–252.

Lacalli, T.C., 2005. Protochordate body plan and the evolutionary role of larvae: old controversies resolved? Canadian Journal of Zoology 83, 216–224.

Lacalli, T.C., Hou, S.F., 1999. A reexamination of the epithelial sensory cells of amphioxus (*Branchiostoma*). Acta Zoologica 80, 125–134.

Lamarck, J.B., 1794. Recherches sur les causes des principaux faits physiques. (Maradan Paris, France).

Lamarck, J.B., 1816. Histoire naturelle des animaux sans vertebres, vol. III. Tuniciers, De'terville, Paris, France.

Langeland, J.A., Kimmel, C.B., 1997. Fishes. In: Gilbert, S.F., Raunio, A.M. (Eds.), Embryology: Constructing the Organism. Sinauer Associates Inc., Mass, USA, pp. 383–407.

Lankester, E.R., 1877. Notes on the embryology and classification of the animal kingdom: comprising a revision of speculations relative to the origin and significance of germ layers. Quarterly Journal of Microscopical Science 17, 399–454.

Lapraz, F., Besnardeau, L., Lepage, T., 2009. Patterning of the dorsal-ventral axis in echinoderms: insights into the evolution of the BMP-chordin signaling network. PLoS Biology 7, e1000248.

Lapraz, F., Haillot, E., Lepage, T., 2015. A deuterostome origin of the Spemann organiser suggested by Nodal and ADMPs functions in Echinoderms. Nature Communications 6, 8434.

Laumer, C.E., Bekkouche, N., Kerbl, A., Goetz, F., Neves, R.C., Sørensen, M.V., Kristensen, R.M., Hejnol, A., Dunn, C.W., Giribet, G., Worsaae, K., 2015. Spiralian phylogeny informs the evolution of microscopic lineages. Current Biology 25, 2000–2006.

Lauri, A., Brunet, T., Handberg-Thorsager, M., Fischer, A.H., Simakov, O., Steinmetz, P.R., Tomer, R., Keller, P.J., Arendt, D., 2014. Development of the annelid axochord: insights into notochord evolution. Science 345, 1365–1368.

Le Douarin, N., 1982. The Neural Crest. Cambridge University Press.

Le Douarin, N., Kalcheim, C., 1999. The Neural Crest, second ed. Cambridge University Press, Cambridge, UK.

Lemaire, P., 2011. Evolutionary crossroads in developmental biology: the tunicates. Development 138, 2143–2152.

Levin, M., 2005. Left-right asymmetry in embryonic development: a comprehensive review. Mechanisms of Development 122, 3–25.

Levin, M., Hashimshony, T., Wagner, F., Yanai, I., 2012. Developmental milestones punctuate gene expression in the *Caenorhabditis* embryo. Developmental Cell 22, 1101–1108.

Liang, D., Wu, R., Geng, J., Wang, C., Zhang, P., 2011. A general scenario of *Hox* gene inventory variation among major sarcopterygian lineages. BMC Evolutionary Biology 11, 25.

Lin, C.Y., Tung, C.H., Yu, J.K., Su, Y.H., 2016. Reproductive periodicity, spawning induction, and larval metamorphosis of the hemichordate acorn worm *Ptychodera flava*. Journal of Experimental Zoology. Part B Molecular and Developmental Evolution 326, 47–60.

Linnaeus, C., 1766–1767. Systema Naturae, twelfth ed. Holmiae, Salvius, Sweden.

Linsenmayer, T.F., Gibney, E., Fitch, J.M., 1986. Embryonic avian cornea contains layers of collagen with greater than average stability. Journal of Cell Biology 103, 1587–1593.

Livingston, B.T., Killian, C.E., Wilt, F., Cameron, A., Landrum, M.J., Ermolaeva, O., Sapojnikov, V., Maglott, D.R., Buchanan, A.M., Ettensohn, C.A., 2006. A genome-wide analysis of biomineralization-related proteins in the sea urchin *Strongylocentrotus purpuratus*. Developmental Biology 300, 335–348.

Lolas, M., Valenzuela, P.D., Tjian, R., Liu, Z., 2014. Charting Brachyury-mediated developmental pathways during early mouse embryogenesis. Proceedings of the National Academy of Sciences of the United States of America 111, 4478–4483.

Lovtrup, S., 1977. The Phylogeny of Vertebrata. John Wiley & Sons, New York.

Lowe, C.J., Clarke, D.N., Medeiros, D.M., Rokhsar, D.S., Gerhart, J., 2015. The deuterostome context of chordate origins. Nature 520, 456–465.

Lowe, C.J., Terasaki, M., Wu, M., Freeman Jr., R.M., Runft, L., Kwan, K., Haigo, S., Aronowicz, J., Lander, E., Gruber, C., Smith, M., Kirschner, M., Gerhart, J., 2006. Dorsoventral patterning in hemichordates: insights into early chordate evolution. PLoS Biology 4, e291.

Lowe, C.J., Wu, M., Salic, A., Evans, L., Lander, E., Stange-Thomann, N., Gruber, C.E., Gerhart, J., Kirschner, M., 2003. Anteroposterior patterning in hemichordates and the origins of the chordate nervous system. Cell 113, 853–865.

Luo, Y.J., Takeuchi, T., Koyanagi, R., Yamada, L., Kanda, M., Khalturina, M., Fujie, M., Yamasaki, S., Endo, K., Satoh, N., 2015. The *Lingula* genome provides insights into brachiopod evolution and the origin of phosphate biomineralization. Nature Communications 6, 8301.

Luttrell, S., Konikoff, C., Byrne, A., Bengtsson, B., Swalla, B.J., 2012. Ptychoderid hemichordate neurulation without a notochord. Integrative and Comparative Biology 52, 829–834.

Mallatt, J., Chen, J., Holland, N.D., 2003. Comment on "A new species of yunnanozoan with implications for deuterostome evolution". Science 300, 1372.

Mansfield, J.H., Haller, E., Holland, N.D., Brent, A.E., 2015. Development of somites and their derivatives in amphioxus, and implications for the evolution of vertebrate somites. Evodevo 6, 21.

Margulis, L., Schwartz, K.V., 1998. Five Kingdoms, third ed. Freeman & Co., New York, NY.

Matsumoto, J., Kumano, G., Nishida, H., 2007. Direct activation by Ets and Zic is required for initial expression of the *Brachyury* gene in the ascidian notochord. Developmental Biology 306, 870–882.

Matthysse, A.G., Deschet, K., Williams, M., Marry, M., White, A.R., Smith, W.C., 2004. A functional cellulose synthase from ascidian epidermis. Proceedings of the National Academy of Sciences of the United States of America 101, 986–991.

Mazet, F., Hutt, J.A., Milloz, J., Millard, J., Graham, A., Shimeld, S.M., 2005. Molecular evidence from *Ciona intestinalis* for the evolutionary origin of vertebrate sensory placodes. Developmental Biology 282, 494–508.

McClay, D.R., 2011. Evolutionary crossroads in developmental biology: sea urchins. Development 138, 2639–2648.

Medeiros, D.M., 2015. Ancient origin for the axochord: a putative notochord homolog. Bioessays 37, 834.

Mehta, T.K., Ravi, V., Yamasaki, S., Lee, A.P., Lian, M.M., Tay, B.H., Tohari, S., Yanai, S., Tay, A., Brenner, S., Venkatesh, B., 2013. Evidence for at least six Hox clusters in the Japanese lamprey (*Lethenteron japonicum*). Proceedings of the National Academy of Sciences of the United States of America 110, 16044–16049.

Meilhac, S.M., Esner, M., Kelly, R.G., Nicolas, J.F., Buckingham, M.E., 2004. The clonal origin of myocardial cells in different regions of the embryonic mouse heart. Developmental Cell 6, 685–698.

Metchnikoff, V., 1881. Über die systematische Stellung von Balanoglossus. Zoologischer Anzeiger 4, 139–157.

Meulemans, D., Bronner-Fraser, M., 2004. Gene-regulatory interactions in neural crest evolution and development. Developmental Cell 7, 291–299.

Mitchell, C.E., Melchin, M.J., Cameron, C.B., Maletz, J., 2013. Phylogenetic analysis reveals that *Rhabdopleura* is an extant graptolite. Lethaia 46, 34–56.

Miyamoto, N., Nakajima, Y., Wada, H., Saito, Y., 2010. Development of the nervous system in the acorn worm *Balanoglossus simodensis*: insights into nervous system evolution. Evolution and Development 12, 416–424.

Miyamoto, N., Wada, H., 2013. Hemichordate neurulation and the origin of the neural tube. Nature Communications 4, 2713.

Mizutani, C.M., Bier, E., 2008. EvoD/Vo: the origins of BMP signalling in the neuroectoderm. Nature Reviews Genetics 9, 663–677.

Morgan, J.L.W., Strumillo, J., Zimmer, J., 2013. Crystallographic snapshot of cellulose synthesis and membrane translocation. Nature 493, 181–186.

Morgan, T., 1894. Development of *Balanoglossus*. Journal of Morphology 9, 1–86.

Morgan, T.H., 1891. The growth and metamorphosis of tornaria. Journal of Morphology 5, 407–458.

Morley, R.H., Lachani, K., Keefe, D., Gilchrist, M.J., Flicek, P., Smith, J.C., Wardle, F.C., 2009. A gene regulatory network directed by zebrafish No tail accounts for its roles in mesoderm formation. Proceedings of the National Academy of Sciences of the United States of America 106, 3829–3834.

Moroz, L.L., Kocot, K.M., Citarella, M.R., Dosung, S., Norekian, T.P., Povolotskaya, I.S., Grigorenko, A.P., Dailey, C., Berezikov, E., Buckley, K.M., Ptitsyn, A., Reshetov, D., Mukherjee, K., Moroz, T.P., Bobkova, Y., Yu, F.H., Kapitonov, V.V., Jurka, J., Bobkov, Y.V., Swore, J.J., Girardo, D.O., Fodor, A., Gusev, F., Sanford, R., Bruders, R., Kittler, E., Mills, C.E., Rast, J.P., Derelle, R., Solovyev, V.V., Kondrashov, F.A., Swalla, B.J., Sweedler, J.V., Rogaev, E.I., Halanych, K.M., Kohn, A.B., 2014. The ctenophore genome and the evolutionary origins of neural systems. Nature 510, 109–114.

Mungpakdee, S., Seo, H.C., Angotzi, A.R., Dong, X., Akalin, A., Chourrout, D., 2008. Differential evolution of the 13 Atlantic salmon Hox clusters. Molecular Biology and Evolution 25, 1333–1343.

Munro, E.M., Odell, G.M., 2002. Polarized basolateral cell motility underlies invagination and convergent extension of the ascidian notochord. Development 129, 13–24.

Murdock, D.J., Dong, X.P., Repetski, J.E., Marone, F., Stampanoni, M., Donoghue, P.C., 2013. The origin of conodonts and of vertebrate mineralized skeletons. Nature 502, 546–549.

Nakano, H., Lundin, K., Bourlat, S.J., Telford, M.J., Funch, P., Nyengaard, J.R., Obst, M., Thorndyke, M.C., 2013. *Xenoturbella bocki* exhibits direct development with similarities to Acoelomorpha. Nature Communications 4, 1537.

Nakashima, K., Nishino, A., Horikawa, Y., Hirose, E., Sugiyama, J., Satoh, N., 2011. The crystalline phase of cellulose changes under developmental control in a marine chordate. Cellular and Molecular Life Sciences 68, 1623–1631.

Nakashima, K., Yamada, L., Satou, Y., Azuma, J., Satoh, N., 2004. The evolutionary origin of animal cellulose synthase. Development Genes and Evolution 214, 81–88.

Nathan, E., Monovich, A., Tirosh-Finkel, L., Harrelson, Z., Rousso, T., Rinon, A., Harel, I., Evans, S.M., Tzahor, E., 2008. The contribution of Islet1-expressing splanchnic mesoderm cells to distinct branchiomeric muscles reveals significant heterogeneity in head muscle development. Development 135, 647–657.

Nibu, Y., José-Edwards, D.S., Di Gregorio, A., 2013. From notochord formation to hereditary chordoma: the many roles of *Brachyury*. BioMed Research International 2013:826435.

Nielsen, C., 1999. Origin of the chordate central nervous system - and the origin of chordates. Development Genes and Evolution 209, 198–205.

Nielsen, C., 2012. Animal Evolution: Interrelationships of the Living Phyla, third ed. Oxford University Press, New York, NY.

Nieuwkoop, P.D., 1967. The "organization centre." II. Field phenomena, their origin and significance. Acta Biotheoretica 17, 151–177.

Nieuwkoop, P.D., 1973. The organization center of the amphibian embryo: its origin, spatial organization, and morphogenetic action. Advances in Morphogenesis 10, 1–39.

Nishida, H., 2008. Development of the appendicularian *Oikopleura dioica*: culture, genome, and cell lineages. Development Growth and Differentiation 50, S239–S256.

Nishida, H., Sawada, K., 2001. Macho-1 encodes a localized mRNA in ascidian eggs that specifies muscle fate during embryogenesis. Nature 409, 724–729.

Nishino, A., Satoh, N., 2001. The simple tail of chordates: phylogenetic significance of appendicularians. Genesis 29, 36–45.

Nohara, M., Nishida, M., Miya, M., Nishikawa, T., 2005. Evolution of the mitochondrial genome in cephalochordata as inferred from complete nucleotide sequences from two *Epigonichthys* species. Journal of Molecular Evolution 60, 526–537.

Nomaksteinsky, M., Röttinger, E., Dufour, H.D., Chettouh, Z., Lowe, C.J., Martindale, M.Q., Brunet, J.F., 2009. Centralization of the deuterostome nervous system predates chordates. Current Biology 19, 1264–1269.

Nübler-Jung, K., Arendt, D., 1996. Enteropneusts and chordate evolution. Current Biology 6, 352–353.

Oda, H., Wada, H., Tagawa, K., Akiyama-Oda, Y., Satoh, N., Humphreys, T., Zhang, S.C., Tsukita, S., 2002. A novel amphioxus cadherin that localizes to epithelial adherens junctions has an unusual domain organization with implications for chordate phylogeny. Evolution and Development 4, 426–434.

Ogasawara, M., Wada, H., Peters, H., Satoh, N., 1999. Developmental expression of *Pax1/9* genes in urochordate and hemichordate gills: insight into function and evolution of the pharyngeal epithelium. Development 126, 2539–2550.

Ohno, S., 1970. Evolution by Gene Duplication. Springer-Verlag, Berlin, Germany.

Onai, T., Yu, J.K., Blitz, I.L., Cho, K.W., Holland, L.Z., 2010. Opposing Nodal/Vg1 and BMP signals mediate axial patterning in embryos of the basal chordate amphioxus. Developmental Biology 344, 377–389.

Onichtchouk, D., Gawantka, V., Dosch, R., Delius, H., Hirschfeld, K., Blumenstock, C., Niehrs, C., 1996. The *Xvent-2* homeobox gene is part of the *BMP-4* signalling pathway controlling dorsoventral patterning of *Xenopus* mesoderm. Development 122, 3045–3053.

Osborn, K.J., Kuhnz, L.A., Priede, I.G., Urata, M., Gebruk, A.V., Holland, N.D., 2012. Diversification of acorn worms (Hemichordata, Enteropneusta) revealed in the deep sea. Proceedings of the Royal Society B: Biological Sciences 279, 1646–1654.

Pancer, Z., Amemiya, C.T., Ehrhardt, G.R., Ceitlin, J., Gartland, G.L., Cooper, M.D., 2004. Somatic diversification of variable lymphocyte receptors in the agnathan sea lamprey. Nature 430, 174–180.

Pani, A.M., Mullarkey, E.E., Aronowicz, J., Assimacopoulos, S., Grove, E.A., Lowe, C.J., 2012. Ancient deuterostome origins of vertebrate brain signalling centres. Nature 483, 289–294.

Panopoulou, G., Hennig, S., Groth, D., Krause, A., Poustka, A.J., Herwig, R., Vingron, M., Lehrach, H., 2003. New evidence for genome-wide duplications at the origin of vertebrates using an amphioxus gene set and completed animal genomes. Genome Research 13, 1056–1066.

Papaioannou, V.E., 2001. T-box genes in development: from hydra to humans. International Review of Cytology 207, 1–70.

Papaioannou, V.E., 2014. The T-box gene family: emerging roles in development, stem cells and cancer. Development 141, 3819–3833.

Pascual-Anaya, J., Adachi, N., Alvarez, S., Kuratani, S., D'Aniello, S., Garcia-Fernàndez, J., 2012. Broken colinearity of the amphioxus *Hox* cluster. Evodevo 3, 28.

Pascual-Anaya, J., D'Aniello, S., Kuratani, S., Garcia-Fernàndez, J., 2013. Evolution of *Hox* gene clusters in deuterostomes. BMC Developmental Biology 13, 26.

Patthey, C., Schlosser, G., Shimeld, S.M., 2014. The evolutionary history of vertebrate cranial placodes–I: cell type evolution. Developmental Biology 389, 82–97.

Paul, C.R.C., Smith, A.B., 1984. The early radiation and phylogeny of echinoderms. Biological Reviews of the Cambridge Philosophical Society 59, 443–481.

Perseke, M., Golombek, A., Schlegel, M., Struck, T.H., 2013. The impact of mitochondrial genome analyses on the understanding of deuterostome phylogeny. Molecular Phylogenetics and Evolution 66, 898–905.

Peter, I.S., Davidson, E.H., 2015. Genomic Control Process: Development and Evolution, first ed. Academic Press, San Diego, CA.

Peterson, K.J., Cameron, R.A., Tagawa, K., Satoh, N., Davidson, E.H., 1999a. A comparative molecular approach to mesodermal patterning in basal deuterostomes: the expression pattern of *Brachyury* in the enteropneust hemichordate *Ptychodera flava*. Development 126, 85–95.

Peterson, K.J., Cotton, J.A., Gehling, J.G., Pisani, D., 2008. The Ediacaran emergence of bilaterians: congruence between the genetic and the geological fossil records. Philosophical Transactions of the Royal Society B-Biological Sciences 363, 1435–1443.

Peterson, K.J., Dietrich, M.R., McPeek, M.A., 2009. MicroRNAs and metazoan macroevolution: insights into canalization, complexity, and the Cambrian explosion. Bioessays 31, 736–747.

Peterson, K.J., Harada, Y., Cameron, R.A., Davidson, E.H., 1999b. Expression pattern of *Brachyury* and *not* in the sea urchin: comparative implications for the origins of mesoderm in the basal deuterostomes. Developmental Biology 207, 419–431.

Peterson, K.J., McPeek, M.A., Evans, D.A.D., 2005. Tempo and mode of early animal evolution: inferences from rocks, Hox, and molecular clocks. Paleobiology 31, 36–55.

Peterson, K.J., Su, Y.H., Arnone, M.I., Swalla, B., King, B.L., 2013. MicroRNAs support the monophyly of enteropneust hemichordates. Journal of Experimental Zoology Part B: Molecular and Developmental Evolution 320, 368–374.

Philippe, H., Brinkmann, H., Copley, R.R., Moroz, L.L., Nakano, H., Poustka, A.J., Wallberg, A., Peterson, K.J., Telford, M.J., 2011. Acoelomorph flatworms are deuterostomes related to *Xenoturbella*. Nature 470, 255–258.

Philippe, H., Derelle, R., Lopez, P., Pick, K., Borchiellini, C., Boury-Esnault, N., Vacelet, J., Renard, E., Houliston, E., Quéinnec, E., Da Silva, C., Wincker, P., Le Guyader, H., Leys, S., Jackson, D.J., Schreiber, F., Erpenbeck, D., Morgenstern, B., Wörheide, G., Manuel, M., 2009. Phylogenomics revives traditional views on deep animal relationships. Current Biology 19, 706–712.

Poe, S., Wake, M.H., 2004. Quantitative tests of general models for the evolution of development. American Naturalist 164, 415–422.

Poss, S.G., Boschung, H.T., 1996. Lancelets (Cephalochordata: Branchiostomatidae): how many species are valid? Israel Journal of Zoology 42 (Suppl.), 13–66.

Putnam, N.H., Srivastava, M., Hellsten, U., Dirks, B., Chapman, J., Salamov, A., Terry, A., Shapiro, H., Lindquist, E., Kapitonov, V.V., Jurka, J., Genikhovich, G., Grigoriev, I.V., Lucas, S.M., Steele, R.E., Finnerty, J.R., Technau, U., Martindale, M.Q., Rokhsar, D.S., 2007. Sea anemone genome reveals ancestral eumetazoan gene repertoire and genomic organization. Science 317, 86–94.

Putnam, N.H., Butts, T., Ferrier, D.E.K., Furlong, R.F., Hellsten, U., Kawashima, T., Robinson-Rechavi, M., Shoguchi, E., Terry, A., Yu, J.K., Benito-Gutiérrez, E., Dubchak, I., Garcia-Fernàndez, J., Gibson-Brown, J.J., Grigoriev, I.V., Horton, A.C., de Jong, P.J., Jurka, J., Kapitonov, V.V., Kohara, Y., Kuroki, Y., Lindquist, E., Lucas, S., Osoegawa, K., Pennacchio, L.A., Salamov, A.A., Satou, Y., Sauka-Spengler, T., Schmutz, J., Shin-I, T., Toyoda, A., Bronner-Fraser, M., Fujiyama, A., Holland, L.Z., Holland, P.W.H., Satoh, N., Rokhsar, D.S., 2008. The amphioxus genome and the evolution of the chordate karyotype. Nature 453, 1064–1071.

Quint, M., Drost, H.G., Gabel, A., Ullrich, K.K., Bönn, M., Grosse, I., 2012. A transcriptomic hourglass in plant embryogenesis. Nature 490, 98–101.

Raff, A., 1996. The Shape of Life: Genes, Development, and the Evolution of Animal Form. University of Chicago Press, Chicago, IL.

Rast, J.P., Cameron, R.A., Poustka, A.J., Davidson, E.H., 2002. Brachyury target genes in the early sea urchin embryo isolated by differential macroarray screening. Developmental Biology 246, 191–208.

Reversade, B., De Robertis, E.M., 2005. Regulation of ADMP and BMP2/4/7 at opposite embryonic poles generates a self-regulating morphogenetic field. Cell 123, 1147–1160.

Richardson, M.K., Hanken, J., Gooneratne, M.L., Pieau, C., Raynaud, A., Selwood, L., Wright, G.M., 1997. There is no highly conserved embryonic stage in the vertebrates: implications for current theories of evolution and development. Anatomy and Embryology 196, 91–106.

Richardson, M.K., Minelli, A., Coates, M., Hanken, J., 1998. Phylotypic stage theory. Trends in Ecology and Evolution 13, 158.

Romer, A.S., 1967. Major steps in vertebrate evolution. Science 158, 1629–1637.

Romer, A.S., 1970. The Vertebrate Body, fourth ed. Sanders & Co., Philadelphia, PA.

Romer, A.S., 1972. The vertebrate as a dual animal-somatic and visceral. Evolutionary Biology 6, 121–156.

Röttinger, E., DuBuc, T.Q., Amiel, A.R., Martindale, M.Q., 2015. Nodal signaling is required for mesodermal and ventral but not for dorsal fates in the indirect developing hemichordate, *Ptychodera flava*. Biology Open 4, 830–842.

Röttinger, E., Lowe, C.J., 2012. Evolutionary crossroads in developmental biology: hemichordates. Development 139, 2463–2475.

Royal Ontario Museum, 2011. The Burgess Shale. Royal Ontario Museum. http://burgess-shale.rom.on.ca/.

Ruppert, E.E., 1997. Cephalochordata (Acrania). In: Harrison, F.W., Ruppert, E.E. (Eds.), Microscopic Anatomy of Invertebrates. Hemichordata, Chaetognatha, and the Invertebrate Chordates, vol. 15. Wiley-Liss, New York, NY, pp. 349–504.

Ruppert, E.E., 2005. Key characters uniting hemichordates and chordates: homologies or homoplasies? Canadian Journal of Zoology 83, 8–23.

Ruppert, E.E., Fox, R.S., Barnes, R.D., 2004. Invertebrate Zoology: A Functional Evolutionary Approach, seventh ed. Thomson-Brooks/Cole, Belmont, CA.

Ryan, J.F., Pang, K., Schnitzler, C.E., Nguyen, A.D., Moreland, R.T., Simmons, D.K., Koch, B.J., Francis, W.R., Havlak, P., Program, N.C.S., Smith, S.A., Putnam, N.H., Haddock, S.H.D., Dunn, C.W., Wolfsberg, T.G., Mullikin, J.C., Martindale, M.Q., Baxevanis, A.D., 2013. The genome of the ctenophore *Mnemiopsis leidyi* and its implications for cell yype evolution. Science 342, 1242592.

Sagane, Y., Zech, K., Bouquet, J.M., Schmid, M., Bal, U., Thompson, E.M., 2010. Functional specialization of cellulose synthase genes of prokaryotic origin in chordate larvaceans. Development 137, 1483–1492.

Salvini-Plawen, L., 1999. On the phylogenetic significance of the neurenteric canal (Chordata). Zoology-Analysis of Complex Systems 102, 175–183.

Sanges, R., Hadzhiev, Y., Gueroult-Bellone, M., Roure, A., Ferg, M., Meola, N., Amore, G., Basu, S., Brown, E.R., De Simone, M., Petrera, F., Licastro, D., Strähle, U., Banfi, S., Lemaire, P., Birney, E., Müller, F., Stupka, E., 2013. Highly conserved elements discovered in vertebrates are present in non-syntenic loci of tunicates, act as enhancers and can be transcribed during development. Nucleic Acids Research 41, 3600–3618.

Sansom, R.S., Gabbott, S.E., Purnell, M.A., 2010. Non-random decay of chordate characters causes bias in fossil interpretation. Nature 463, 797–800.

Santagati, F., Gerber, J.K., Blusch, J.H., Kokubu, C., Peters, H., Adamski, J., Werner, T., Balling, R., Imai, K., 2001. Comparative analysis of the genomic organization of *Pax9* and its conserved physical association with *Nkx2-9* in the human, mouse, and pufferfish genomes. Mammalian Genome 12, 232–237.

Santos, R., Haesaerts, D., Jangoux, M., Flammang, P., 2005. The tube feet of sea urchins and sea stars contain functionally different mutable collagenous tissues. Journal of Experimental Biology 208, 2277–2288.

Sasakura, Y., Nakashima, K., Awazu, S., Matsuoka, T., Nakayama, A., Azuma, J., Satoh, N., 2005. Transposon-mediated insertional mutagenesis revealed the functions of animal cellulose synthase in the ascidian *Ciona intestinalis*. Proceedings of the National Academy of Sciences of the United States of America 102, 15134–15139.

Sato, A., Rickards, B., Holland, P.W.H., 2008. The origins of graptolites and other pterobranchs: a journey from 'Polyzoa'. Lethaia 41, 303–316.

Satoh, N., 1994. Developmental Biology of Ascidians. Cambridge University Press, NY, USA.

Satoh, N., 2003. The ascidian tadpole larva: comparative molecular development and genomics. Nature Reviews Genetics 4, 285–295 3.

Satoh, N., 2008. An aboral-dorsalization hypothesis for chordate origin. Genesis 46, 614–622.

Satoh, N., 2009. An advanced filter-feeder hypothesis for urochordate evolution. Zoological Science 26, 97–111.

Satoh, N., 2014. Developmental Genomics of Ascidians. Wiley-Blackwell, New Jersey, USA.

Satoh, N., Rokhsar, D., Nishikawa, T., 2014a. Chordate evolution and the three-phylum system. Proceedings of the Royal Society B: Biological Sciences 281, 20141729.

Satoh, N., Tagawa, K., Lowe, C.J., Yu, J.K., Kawashima, T., Takahashi, H., Ogasawara, M., Kirschner, M., Hisata, K., Su, Y.H., Gerhart, J., 2014b. On a possible evolutionary link of the stomochord of hemichordates to pharyngeal organs of chordates. Genesis 52, 925–934.

Satoh, N., Tagawa, K., Takahashi, H., 2012. How was the notochord born? Evolution and Development 14, 56–75.

Satou, Y., Mineta, K., Ogasawara, M., Sasakura, Y., Shoguchi, E., Ueno, K., Yamada, L., Matsumoto, J., Wasserscheid, J., Dewar, K., Wiley, G.B., Macmil, S.L., Roe, B.A., Zeller, R.W., Hastings, K.E.M., Lemaire, P., Lindquist, E., Endo, T., Hotta, K., Inaba, K., 2008. Improved genome assembly and evidence-based global gene model set for the chordate *Ciona intestinalis*: new insight into intron and operon populations. Genome Biology 9, R152.

Schaeffer, B., 1987. Deuterostome monophyly and phylogeny. Evolutionary Biology 21, 179–235.

Schlosser, G., Patthey, C., Shimeld, S.M., 2014. The evolutionary history of vertebrate cranial placodes II. Evolution of ectodermal patterning. Developmental Biology 389, 98–119.

Schmid, T.M., Bonen, D.K., Luchene, L., Linsenmayer, T.F., 1991. Late events in chondrocyte differentiation: hypertrophy, type X collagen synthesis and matrix calcification. In Vivo 5, 533–540 (Athens, Greece).

Schohl, A., Fagotto, F., 2002. β-catenin, MAPK and Smad signaling during early *Xenopus* development. Development 129, 37–52.

Schulte-Merker, S., van Eeden, F.J., Halpern, M.E., Kimmel, C.B., Nüsslein-Volhard, C., 1994. *No tail* (*ntl*) is the zebrafish homologue of the mouse *T* (*Brachyury*) gene. Development 120, 1009–1015.

Sea Urchin Genome Sequencing Consortium, Sodergren, E., Weinstock, G.M., Davidson, E.H., Cameron, R.A., Gibbs, R.A., Angerer, R.C., Angerer, L.M., Arnone, M.I., Burgess, D.R., Burke, R.D., Coffman, J.A., Dean, M., Elphick, M.R., Ettensohn, C.A., Foltz, K.R., Hamdoun, A., Hynes, R.O., Klein, W.H., Marzluff, W., McClay, D.R., Morris, R.L., Mushegian, A., Rast, J.P., Smith, L.C., Thorndyke, M.C., Vacquier, V.D., Wessel, G.M., Wray, G., Zhang, L., Elsik, C.G., Ermolaeva, O., Hlavina, W., Hofmann, G., Kitts, P., Landrum, M.J., Mackey, A.J., Maglott, D., Panopoulou, G., Poustka, A.J., Pruitt, K., Sapojnikov, V., Song, X., Souvorov, A., Solovyev, V., Wei, Z., Whittaker, C.A., Worley, K., Durbin, K.J., Shen, Y., Fedrigo, O., Garfield, D., Haygood, R., Primus, A., Satija, R., Severson, T., Gonzalez-Garay, M.L., Jackson, A.R., Milosavljevic, A., Tong, M., Killian, C.E., Livingston, B.T., Wilt, F.H., Adams, N., Bellé, R., Carbonneau, S., Cheung, R., Cormier, P., Cosson, B., Croce, J., Fernandez-Guerra, A., Geneviève, A.M., Goel, M., Kelkar, H., Morales, J., Mulner-Lorillon, O., Robertson, A.J., Goldstone, J.V., Cole, B., Epel, D., Gold, B., Hahn, M.E., Howard-Ashby, M., Scally, M., Stegeman, J.J., Allgood, E.L., Cool, J., Judkins, K.M., McCafferty, S.S., Musante, A.M., Obar, R.A., Rawson, A.P., Rossetti, B.J., Gibbons, I.R., Hoffman, M.P., Leone, A., Istrail, S., Materna, S.C., Samanta, M.P., Stolc, V., Tongprasit, W., Tu, Q., Bergeron, K.F., Brandhorst, B.P., Whittle, J., Berney, K., Bottjer, D.J., Calestani, C., Peterson, K., Chow, E., Yuan, Q.A., Elhaik, E., Graur, D., Reese, J.T., Bosdet, I., Heesun, S., Marra, M.A., Schein, J., Anderson, M.K., Brockton, V., Buckley, K.M., Cohen, A.H., Fugmann, S.D., Hibino, T., Loza-Coll, M., Majeske, A.J., Messier, C., Nair, S.V., Pancer, Z., Terwilliger, D.P., Agca, C., Arboleda, E., Chen, N., Churcher, A.M., Hallböök, F., Humphrey, G.W., Idris, M.M., Kiyama, T., Liang, S., Mellott, D., Mu, X., Murray, G., Olinski, R.P., Raible, F., Rowe, M., Taylor, J.S., Tessmar-Raible, K., Wang, D., Wilson, K.H., Yaguchi, S., Gaasterland, T., Galindo, B.E., Gunaratne, H.J., Juliano, C., Kinukawa, M., Moy, G.W., Neill, A.T., Nomura, M., Raisch, M., Reade, A., Roux, M.M., Song, J.L., Su, Y.H., Townley, I.K., Voronina, E., Wong, J.L., Amore, G., Branno, M., Brown, E.R., Cavalieri, V., Duboc, V., Duloquin, L., Flytzanis, C., Gache, C., Lapraz, F., Lepage, T., Locascio, A., Martinez, P., Matassi, G., Matranga, V., Range, R., Rizzo, F., Röttinger, E., Beane, W., Bradham, C., Byrum, C., Glenn, T., Hussain, S., Manning, G., Miranda, E., Thomason, R., Walton, K., Wikramanayke, A., Wu, S.Y., Xu, R., Brown, C.T., Chen, L., Gray, R.F., Lee, P.Y., Nam, J., Oliveri, P., Smith, J., Muzny, D., Bell, S., Chacko, J., Cree, A., Curry, S., Davis, C., Dinh, H., Dugan-Rocha, S., Fowler, J., Gill, R., Hamilton, C., Hernandez, J., Hines, S., Hume, J., Jackson, L., Jolivet, A., Kovar, C., Lee, S., Lewis, L., Miner, G., Morgan, M., Nazareth, L.V., Okwuonu, G., Parker, D., Pu, L.L., Thorn, R., Wright, R., 2006. The genome of the sea urchin *Strongylocentrotus purpuratus*. Science 314, 941–952.

Seo, H.C., Edvardsen, R.B., Maeland, A.D., Bjordal, M., Jensen, M.F., Hansen, A., Flaat, M., Weissenbach, J., Lehrach, H., Wincker, P., Reinhardt, R., Chourrout, D., 2004. *Hox* cluster disintegration with persistent anteroposterior order of expression in *Oikopleura dioica*. Nature 431, 67–71.

Shigetani, Y., Sugahara, F., Kuratani, S., 2005. A new evolutionary scenario for the vertebrate jaw. Bioessays 27, 331–338.

Shimeld, S.M., Holland, P.W.H., 2000. Vertebrate innovations. Proceedings of the National Academy of Sciences of the United States of America 97, 4449–4452.

Shoguchi, E., Hamaguchi, M., Satoh, N., 2008. Genome-wide network of regulatory genes for construction of a chordate embryo. Developmental Biology 316, 498–509.

Showell, C., Binder, O., Conlon, F.L., 2004. T-box genes in early embryogenesis. Developmental Dynamics 229, 201–218.

Shu, D.G., Chen, L., Han, J., Zhang, X.L., 2001. An early Cambrian tunicate from China. Nature 411, 472–473.

Shu, D.G., Conway-Morris, S., 2003. Response to comment on "A new species of yunnanozoan with implications for deuterostome evolution". Science 300, 1372.

Shu, D.G., Luo, H.L., Morris, S.C., Zhang, X.L., Hu, S.X., Chen, L., Han, J., Zhu, M., Li, Y., Chen, L.Z., 1999. Lower Cambrian vertebrates from South China. Nature 402, 42–46.

Shu, D.G., Morris, S.C., Han, J., Zhang, Z.F., Yasui, K., Janvier, P., Chen, L., Zhang, X.L., Liu, J.N., Li, Y., Liu, H.Q., 2003. Head and backbone of the early Cambrian vertebrate *Haikouichthys*. Nature 421, 526–529.

Shu, D.G., Morris, S.C., Zhang, X.L., 1996. A *Pikaia*-like chordate from the lower Cambrian of China. Nature 384, 157–158.

Simakov, O., Kawashima, T., Marlétaz, F., Jenkins, J., Koyanagi, R., Mitros, T., Hisata, K., Bredeson, J., Shoguchi, E., Gyoja, F., Yue, J.X., Chen, Y.C., Freeman, R.M., Sasaki, A., Hikosaka-Katayama, T., Sato, A., Fujie, M., Baughman, K.W., Levine, J., Gonzalez, P., Cameron, C., Fritzenwanker, J.H., Pani, A.M., Goto, H., Kanda, M., Arakaki, N., Yamasaki, S., Qu, J., Cree, A., Ding, Y., Dinh, H.H., Dugan, S., Holder, M., Jhangiani, S.N., Kovar, C.L., Lee, S.L., Lewis, L.R., Morton, D., Nazareth, L.V., Okwuonu, G., Santibanez, J., Chen, R., Richards, S., Muzny, D.M., Gillis, A., Peshkin, L., Wu, M., Humphreys, T., Su, Y.H., Putnam, N.H., Schmutz, J., Fujiyama, A., Yu, J.K., Tagawa, K., Worley, K.C., Gibbs, R.A., Kirschner, M.W., Lowe, C.J., Satoh, N., Rokhsar, D.S., Gerhart, J., 2015. Hemichordate genomes and deuterostome origins. Nature 527, 459–465.

Simakov, O., Marletaz, F., Cho, S.J., Edsinger-Gonzales, E., Havlak, P., Hellsten, U., Kuo, D.H., Larsson, T., Lv, J., Arendt, D., Savage, R., Osoegawa, K., de Jong, P., Grimwood, J., Chapman, J.A., Shapiro, H., Aerts, A., Otillar, R.P., Terry, A.Y., Boore, J.L., Grigoriev, I.V., Lindberg, D.R., Seaver, E.C., Weisblat, D.A., Putnam, N.H., Rokhsar, D.S., 2013. Insights into bilaterian evolution from three spiralian genomes. Nature 493, 526–531.

Simões-Costa, M., Bronner, M.E., 2015. Establishing neural crest identity: a gene regulatory recipe. Development 142, 242–257.

Slack, J.M.W., Holland, P.W.H., Graham, C.F., 1993. The zootype and the phylotypic stage. Nature 361, 490–492.

Smith, A.B., 2005. The pre-radial history of echinoderms. Geological Journal 40, 255–280.

Smith, A.B., 2008. Deuterostomes in a twist: the origins of a radical new body plan. Evolution and Development 10, 493–503.

Smith, J.J., Antonacci, F., Eichler, E.E., Amemiya, C.T., 2009. Programmed loss of millions of base pairs from a vertebrate genome. Proceedings of the National Academy of Sciences of the United States of America 106, 11212–11217.

Smith, J.J., Kuraku, S., Holt, C., Sauka-Spengler, T., Jiang, N., Campbell, M.S., Yandell, M.D., Manousaki, T., Meyer, A., Bloom, O.E., Morgan, J.R., Buxbaum, J.D., Sachidanandam, R., Sims, C., Garruss, A.S., Cook, M., Krumlauf, R., Wiedemann, L.M., Sower, S.A., Decatur, W.A., Hall, J.A., Amemiya, C.T., Saha, N.R., Buckley, K.M., Rast, J.P., Das, S., Hirano, M., McCurley, N., Guo, P., Rohner, N., Tabin, C.J., Piccinelli, P., Elgar, G., Ruffier, M., Aken, B.L., Searle, S.M.J., Muffato, M., Pignatelli, M., Herrero, J., Jones, M., Brown, C.T., Chung-Davidson, Y.W., Nanlohy, K.G., Libants, S.V., Yeh, C.Y., McCauley, D.W., Langeland, J.A., Pancer, Z., Fritzsch, B., de Jong, P.J., Zhu, B.L., Fulton, L.L., Theising, B., Flicek, P., Bronner, M.E., Warren, W.C., Clifton, S.W., Wilson, R.K., Li, W.M., 2013. Sequencing of the sea lamprey (*Petromyzon marinus*) genome provides insights into vertebrate evolution. Nature Genetics 45, 415–421.

Smith, M.S., Turner, F.R., Raff, R.A., 2008. Nodal expression and heterochrony in the evolution of dorsal-ventral and left-right axes formation in the direct-developing sea urchin Heliocidaris erythrogramma. Journal of Experimental Zoology Part B: Molecular and Developmental Evolution 310, 609–622.

Smits, P., Lefebvre, V., 2003. *Sox5* and *Sox6* are required for notochord extracellular matrix sheath formation, notochord cell survival and development of the nucleus pulposus of intervertebral discs. Development 130, 1135–1148.

Sodergren, E., Shen, Y., Song, X., Zhang, L., Gibbs, R.A., Weinstock, G.M., 2006. Shedding genomic light on Aristotle's lantern. Developmental Biology 300, 2–8.

Solursh, M., Jensen, K.L., Reiter, R.S., Schmid, T.M., Linsenmayer, T.F., 1986. Environmental regulation of type X collagen production by cultures of limb mesenchyme, mesectoderm, and sternal chondrocytes. Developmental Biology 117, 90–101.

Somorjai, I., Bertrand, S., Camasses, A., Haguenauer, A., Escriva, H., 2008. Evidence for stasis and not genetic piracy in developmental expression patterns of *Branchiostoma lanceolatum* and *Branchiostoma floridae*, two amphioxus species that have evolved independently over the course of 200 Myr. Development Genes and Evolution 218, 703–713.

Soshnikova, N., 2014. Hox genes regulation in vertebrates. Developmental Dynamics 243, 49–58.

Spagnuolo, A., Ristoratore, F., Di Gregorio, A., Aniello, F., Branno, M., Di Lauro, R., 2003. Unusual number and genomic organization of *Hox* genes in the tunicate *Ciona intestinalis*. Gene 309, 71–79.

Spemann, H., Mangold, H., 1924. Induction of embryonic primordia by implantation of organizers from a different species. Roux' Archiv für Entwicklungsmechanik der Organismen 100, 599–638.

Stach, T., 2008. Chordate phylogeny and evolution: a not so simple three-taxon problem. Journal of Zoology 276, 117–141.

Stach, T., Turbeville, J.M., 2002. Phylogeny of Tunicata inferred from molecular and morphological characters. Molecular Phylogenetics and Evolution 25, 408–428.

Stach, T., Winter, J., Bouquet, J.M., Chourrout, D., Schnabel, R., 2008. Embryology of a planktonic tunicate reveals traces of sessility. Proceedings of the National Academy of Sciences of the United States of America 105, 7229–7234.

Stadler, P.F., Fried, C., Prohaska, S.J., Bailey, W.J., Misof, B.Y., Ruddle, F.H., Wagner, G.P., 2004. Evidence for independent *Hox* gene duplications in the hagfish lineage: a PCR-based gene inventory of *Eptatretus stoutii*. Molecular Phylogenetics and Evolution 32, 686–694.

Stemple, D.L., 2005. Structure and function of the notochord: an essential organ for chordate development. Development 132, 2503–2512.

Stern, C.D., 2004. Gastrulation: From Cells to Embryo. Cold Spring Harbor Laboratory Press, NY, US.

Steventon, B., Mayor, R., Streit, A., 2014. Neural crest and placode interaction during the development of the cranial sensory system. Developmental Biology 389, 28–38.

Suzuki, A., Ueno, N., HemmatiBrivanlou, A., 1997. *Xenopus msx1* mediates epidermal induction and neural inhibition by BMP4. Development 124, 3037–3044.

Suzuki, M.M., Satoh, N., 2000. Genes expressed in the amphioxus notochord revealed by EST analysis. Developmental Biology 224, 168–177.

Swalla, B.J., 2006. Building divergent body plans with similar genetic pathways. Heredity 97, 235–243.

Swalla, B.J., Cameron, C.B., Corley, L.S., Garey, J.R., 2000. Urochordates are monophyletic within the deuterostomes. Systematic Biology 49, 52–64.

Swalla, B.J., Smith, A.B., 2008. Deciphering deuterostome phylogeny: molecular, morphological and palaeontological perspectives. Philosophical Transactions of the Royal Society B: Biological Sciences 363, 1557–1568.

Tagawa, K., Humphreys, T., Satoh, N., 1998. Novel pattern of *Brachyury* gene expression in hemichordate embryos. Mechanisms of Development 75, 139–143.

Takacs, C.M., Moy, V.N., Peterson, K.J., 2002. Testing putative hemichordate homologues of the chordate dorsal nervous system and endostyle: expression of *NK2.1* (*TTF-1*) in the acorn worm *Ptychodera flava* (Hemichordata, Ptychoderidae). Evolution and Development 4, 405–417.

Takahashi, H., Hotta, K., Erives, A., Di Gregorio, A., Zeller, R.W., Levine, M., Satoh, N., 1999. Brachyury downstream notochord differentiation in the ascidian embryo. Genes and Development 13, 1519–1523.

Tao, J., Kuliyev, E., Wang, X., Li, X., Wilanowski, T., Jane, S.M., Mead, P.E., Cunningham, J.M., 2005. BMP4-dependent expression of *Xenopus* Grainyhead-like 1 is essential for epidermal differentiation. Development 132, 1021–1034.

Tarver, J.E., Sperling, E.A., Nailor, A., Heimberg, A.M., Robinson, J.M., King, B.L., Pisani, D., Donoghue, P.C.J., Peterson, K.J., 2013. miRNAs: small genes with big potential in metazoan phylogenetics. Molecular Biology and Evolution 30, 2369–2382.

Telford, M.J., 2008. Xenoturbellida: the fourth deuterostome phylum and the diet of worms. Genesis 46, 580–586.

Terakita, A., 2005. The opsins. Genome Biology 6, 213.

Tokioka, T., 1971. Phylogenetic speculation of the Tunicata. Publications of the Seto Marine Biological Laboratory 19, 43–63.

Tuazon, F.B., Mullins, M.C., 2015. Temporally coordinated signals progressively pattern the anteroposterior and dorsoventral body axes. Seminars in Cell and Developmental Biology 42, 118–133.

Turbeville, J.M., Schulz, J.R., Raff, R.A., 1994. Deuterostome phylogeny and the sister group of the chordates – evidence from molecules and morphology. Molecular Biology and Evolution 11, 648–655.

Turon, X., López-Legentil, S., 2004. Ascidian molecular phylogeny inferred from mtDNA data with emphasis on the Aplousobranchiata. Molecular Phylogenetics and Evolution 33, 309–320.

Urano, A., Suzuki, M.M., Zhang, P., Satoh, N., Satoh, G., 2003. Expression of muscle-related genes and two MyoD genes during amphioxus notochord development. Evolution and Development 5, 447–458.

Vandekerckhove, J., Weber, K., 1979. The complete amino-acid sequence of actins from bovine aorta, bovine heart, bovine fast skeletal muscle, and rabbit slow skeletal muscle. A protein-chemical analysis of muscle actin differentiation. Differentiation 14, 123–133.

Vandekerckhove, J., Weber, K., 1984. Chordate muscle actins differ distinctly from invertebrate muscle actins – the evolution of the different vertebrate muscle actins. Journal of Molecular Biology 179, 391–413.

Venkatesh, B., Lee, A.P., Ravi, V., Maurya, A.K., Lian, M.M., Swann, J.B., Ohta, Y., Flajnik, M.F., Sutoh, Y., Kasahara, M., Hoon, S., Gangu, V., Roy, S.W., Irimia, M., Korzh, V., Kondrychyn, I., Lim, Z.W., Tay, B.H., Tohari, S., Kong, K.W., Ho, S.F., Lorente-Galdos, B., Quilez, J., Marques-Bonet, T., Raney, B.J., Ingham, P.W., Tay, A., Hillier, L.W., Minx, P., Boehm, T., Wilson, R.K., Brenner, S., Warren, W.C., 2014. Elephant shark genome provides unique insights into gnathostome evolution. Nature 505, 174–179.

von Baer, K.E., 1828. Über Entwickelungsgeschichte der Thiere : Beobachtung und Reflexion. Gebrüdern Bornträger, Königsberg.

Voskoboynik, A., Newman, A.M., Corey, D.M., Sahoo, D., Pushkarev, D., Neff, N.F., Passarelli, B., Koh, W., Ishizuka, K.J., Palmeri, K.J., Dimov, I.K., Keasar, C., Fan, H.C., Mantalas, G.L., Sinha, R., Penland, L., Quake, S.R., Weissman, I.L., 2013. Identification of a colonial chordate histocompatibility gene. Science 341, 384–387.

Wada, H., Makabe, K.W., Nakauchi, M., Satoh, N., 1992. Phylogenetic-relationships between solitary and colonial ascidians, as inferred from the sequence of the central region of their respective 18S rDNAs. Biological Bulletin 183, 448–455.

Wada, H., Saiga, H., Satoh, N., Holland, P.W.H., 1998. Tripartite organization of the ancestral chordate brain and the antiquity of placodes: insights from ascidian *Pax-2/5/8*, *Hox* and *Otx* genes. Development 125, 1113–1122.

Wada, H., Satoh, N., 1994. Details of the evolutionary history from invertebrates to vertebrates, as deduced from the sequences of 18S rDNA. Proceedings of the National Academy of Sciences of the United States of America 91, 1801–1804.

Walcott, C., 1911. Cambrian geology and paleontology II. Middle Cambrian annelids. Smithsonian Miscellaneous Collections 57, 109–145.

Wang, Z., Pascual-Anaya, J., Zadissa, A., Li, W.Q., Niimura, Y., Huang, Z.Y., Li, C.Y., White, S., Xiong, Z.Q., Fang, D.M., Wang, B., Ming, Y., Chen, Y., Zheng, Y., Kuraku, S., Pignatelli, M., Herrero, J., Beal, K., Nozawa, M., Li, Q.Y., Wang, J., Zhang, H.Y., Yu, L.L., Shigenobu, S., Wang, J.Y., Liu, J.N., Flicek, P., Searle, S., Wang, J., Kuratani, S., Yin, Y., Aken, B., Zhang, G.J., Irie, N., 2013. The draft genomes of soft-shell turtle and green sea turtle yield insights into the development and evolution of the turtle-specific body plan. Nature Genetics 45, 701–706.

Weaver, C., Kimelman, D., 2004. Move it or lose it: axis specification in *Xenopus*. Development 131, 3491–3499.

Wellik, D.M., Capecchi, M.R., 2003. *Hox10* and *Hox11* genes are required to globally pattern the mammalian skeleton. Science 301, 363–367.

Whittaker, J.R., 1997. Cephalochordates, the lancelets. In: Gilbert, S.F., Raunio, A.M. (Eds.), Embryology: Constructing the Organism. Sinauer Associates Inc., Mass, USA, pp. 365–381.

Wikramanayake, A.H., Hong, M., Lee, P.N., Pang, K., Byrum, C.A., Bince, J.M., Xu, R.H., Martindale, M.Q., 2003. An ancient role for nuclear β-catenin in the evolution of axial polarity and germ layer segregation. Nature 426, 446–450.

Wilkinson, D.G., Bhatt, S., Herrmann, B.G., 1990. Expression pattern of the mouse *T*-gene and tts role in mesoderm formation. Nature 343, 657–659.

Willmer, P., 1990. Invertebrate Relationships: Patterns in Animal Evolution. Cambridge University Press, Cambridge.

Wolpert, L., Tickle, C., 2011. Principles of Development, fourth ed. Oxford University Press, Oxford.

Wray, G.A., 1997. Echinoderms. In: Gilbert, S.F., Raunio, A.M. (Eds.), Embryology: Constructing the Organism. Sinauer Associates Inc., Mass, USA, pp. 309–329.

Wray, G.A., Levinton, J.S., Shapiro, L.H., 1996. Molecular evidence for deep Precambrian divergences among metazoan phyla. Science 274, 568–573.

Wu, H.R., Chen, Y.T., Su, Y.H., Luo, Y.J., Holland, L.Z., Yu, J.K., 2011. Asymmetric localization of germline markers *Vasa* and *Nanos* during early development in the amphioxus *Branchiostoma floridae*. Developmental Biology 353, 147–159.

Yarrell, W., 1836. A History of British Fishes. John Van Voorst, London, UK.

Yasuo, H., Satoh, N., 1993. Function of vertebrate *T* gene. Nature 364, 582–583.

Yokobori, S., Oshima, T., Wada, H., 2005. Complete nucleotide sequence of the mitochondrial genome of *Doliolum nationalis* with implications for evolution of urochordates. Molecular Phylogenetics and Evolution 34, 273–283.

Yu, J.K., 2010. The evolutionary origin of the vertebrate neural crest and its developmental gene regulatory network – insights from amphioxus. Zoology (Jena) 113, 1–9.

Yu, J.K., Satou, Y., Holland, N.D., Shin-I, T., Kohara, Y., Satoh, N., Bronner-Fraser, M., Holland, L.Z., 2007. Axial patterning in cephalochordates and the evolution of the organizer. Nature 445, 613–617.

Zeng, L.Y., Swalla, B.J., 2005. Molecular phylogeny of the protochordates: chordate evolution. Canadian Journal of Zoology-Revue Canadienne De Zoologie 83, 24–33.

Index